Six Sigma *for*
Medical Device Design

Six Sigma *for*
Medical Device Design

Jose Justiniano
Venky Gopalaswamy

CRC PRESS

Boca Raton   London   New York   Washington, D.C.

Library of Congress Cataloging-in-Publication Data

Justiniano, Jose M.
    Six Sigma for medical device design / Jose Justiniano and Venky Gopalaswamy.
      p. cm.
    Includes bibliographical references and index.
    ISBN 0-8493-2105-0 (alk. paper)
      1. Medical instruments and apparatus—Standards. 2. Six sigma (Quality control standard) I. Gopalaswamy, Venky. II. Title.

R856.6.J876 2004
610'.28'4--dc22
                                                   2004051971

Visit the CRC Press Web site at www.crcpress.com

© 2005 by CRC Press

No claim to original U.S. Government works
International Standard Book Number 0-8493-2105-0
Library of Congress Card Number 2004051971
Printed in the United States of America 1 2 3 4 5 6 7 8 9 0
Printed on acid-free paper

Dedication

To my family and friends

VG

To my family and friends, and to those who humbly do the work and know what they are doing.

JMJ

Preface

We believe that any manager at any level who wants to improve his or her chances of launching a medical device with distinctive features of innovation and as fast as possible will benefit from reading this text. It presents a simplistic yet practical view of Design for Six Sigma as the management philosophy, and how it can link the FDA's Design Control requirements of the Quality System Regulation and ISO 13485 with today's competitive global business scenarios. What can be better than a way to comply with regulations and boost business growth and profits? Throughout this book, the authors combine their years of experience in dealing with Six Sigma implementation while attaining regulatory compliance in the medical device field. The practicality of what is written is addressed by providing the "how to" implementation while dispelling commonly held myths concerning Six Sigma deployment and adoption.

In general, Six Sigma and Design for Six Sigma programs will bring top management to a plane of reality where the implementation of the business strategies really takes place. Design for Six Sigma will also help those in the technical and scientific areas of the company to understand the realities of business and markets, while the tools will enable optimization of the product realization process. Bottom line, it sounds like Six Sigma's main contribution is rooted in knowledge-based leadership strengthened by better alignment between management layers, organizations, and the actual doers in the product development organization and all their interfaces.

About the Authors

Jose M. Justiniano has 19 years of experience in quality engineering, reliability, and technical product development management with several leading medical device companies. He brings in-depth expertise in validations of test methods, design, and manufacturing processes for medical devices and pharmaceuticals. He is a seasoned trainer specializing in the design, deployment, and training of quality programs such as the Quality Systems Regulation from the FDA and concepts such as Six Sigma. He has trained more than 4000 vice presidents, directors, managers, life sciences scientists, engineers, and FDA inspectors. Justiniano's teachings have been adopted in the design and manufacturing of immunoassay diagnostic reagents, intra-ocular and contact lenses, endo-surgical equipment, pacemakers and defibrillators, and many other kinds of medical and pharmaceutical products. In 1990, he pioneered the adoption of Six Sigma concepts for medical devices and pharmaceuticals, becoming a trainer and management champion of green and black belts.

Justiniano completed his M.B.A. at the University of Cincinnati and his M.S. in Industrial Engineering and Operations Research at Cornell University. His undergradute degree is in Industrial Engineering from the University of Puerto Rico. He learned the original concepts of Six Sigma directly from Genichi Taguchi and Mikel Harry. He has been recognized by the American Society for Quality (ASQ) as a Certified Quality Engineer (CQE), Certified Reliability Engineer (CRE), Certified Six Sigma Black Belt, and Certified Quality Auditor. He has also attained ISO 9000 Lead Assessor certification by the Institute of Quality Assurance. Justiniano is the co-author of *Practical Design Control Implementation for Medical Devices*, the first book dedicated to medical devices and reliability.

Dr. Venky Gopalaswamy is an executive director in a Fortune 100 company. In his role, he supports Design for Six Sigma (DFSS) development and deployment to enable successful new product, service, and process design and development. He draws upon his extensive experience in product and process development, quality, and project management to develop deployment strategies that leverage organizational strengths and navigate "organizational realities." Dr. Gopalaswamy has coached, mentored, and trained many company associates throughout their DFSS journey. His career spans safety-critical product and process technology development and quality and reliability engineering roles in medical device, high-tech, and automotive industries.

Dr. Gopalaswamy is a company-certified Six Sigma Black Belt. He earned his Ph.D. and M.S. degrees in Industrial and Systems Engineering from the University of Illinois and his undergraduate degree in Mechanical Engineering from Anna University, India. He is also an ASQ Certified Quality Manager (CQM), CRE, and CQE. He has many publications and presentations to his credit in professional journals and conferences. He co-authored the book *Practical Design Control Implementation for Medical Devices*.

Contents

chapter one

Regulation, business, and Six Sigma

The medical device industry (MDI) has grown tremendously over the past few years. Many new and complex devices have been released to the market, and this is just the beginning of a new era of innovation in the healthcare industry. It is becoming common knowledge that the complexity of applications and technologies in the medical device industry is enormous. Today's medical devices range from simple hand-held tools to complex computer-controlled surgical instruments, from implantable screws to artificial organs, from blood-glucose test strips to diagnostic imaging systems and laboratory test equipment. Now we even see hybrid devices or combination products such as combinations of device and pharmaceuticals (e.g., coated stents), products aimed at drug delivery (e.g., photodynamic therapy, which uses a light source to activate a photosensitive drug), and others requiring embedded software to function correctly (e.g., the "camera-in-a-pill" for use in diagnosing bleeding in hard-to-image sections of the intestines). Energy sources such as lasers and high-intensity ultrasound can pass safely through the body, affecting only targeted tissue.

Like in the high-tech, aerospace, and defense industries, the new technologies being assimilated by medical device manufacturers need enhanced methods and control systems for achieving organizational objectives. Example of such objectives are: effective quality planning, quality control, increased customer satisfaction, reduced costs, effective compliance with regulations, shortened product development cycle, quick transfer to manufacturing, and continuous improvement. The methods to deal with quality and regulatory compliance have typically been associated with conventional quality assurance (e.g., inspections, quality audits, humongous documentation) and classical

Quality Engineering (QE) functions (e.g., testing, sampling plans, process capability, Statistical Process Control [SPC], Design of Experiments [DOE], prediction equations).

During the early days of the medical device industry, only very basic quality engineering tools were used in design and development, if any were used at all. In fact, most quality engineering functions were limited to defining acceptance-sampling plans, documenting equipment qualifications, and "approving product/batch release forms" after design transfer (also known as technology transfer in some companies). All these activities could be separated from product design and development and be seen as quality assurance for manufacturing operations.

Today, most companies have realized that quality and reliability are defined from the start of the product development life cycle. In fact, there are even manufacturers that think of quality and reliability from the conception of the product idea. From the regulatory point of view, both the Quality Systems Regulations (QSRs) from the Food and Drug Administration (FDA) and International Standards Organization (ISO) 13485 (2003) impose design controls onto device manufacturers. Besides these, companies that seek to gain some competitive advantage may use quality, reliability, and regulatory compliance as a means for product and company differentiation.

The transformation

Today, there is a transformation in all industries. This transformation began in the late 1970s and is ongoing. It can be described by the typical business jargon we hear these days, such as supply chain, value creation, liquidity, cash flows, profitability, grassroots, minimizing capital investment, vertical and horizontal integrations, acquisitions, technical ladders for engineers and scientists, strategic alliances, diversity, virtual company, global markets, and Six Sigma. The current economic recession that started in the year 2000 and the inevitable increase in the cost of health care will nurture this transformation for the medical device industry. Very simply, large companies making a good profit in a commodities market do not have a wealthy future. At the beginning of year 2000, 75 percent of the 100 largest U.S. companies in the last 70 years were no longer on this list. It is said that the management of such companies failed to recognize the need to reflect the changing environment in their business strategies. Six Sigma is one such strategy, and it is a main element of the aforementioned transformation. We see the adoption of a Six Sigma philosophy by the pharmaceutical and medical device industry as unavoidable.

The tools and the methods of Six Sigma

Over the past decade, medical device companies have begun to use more advanced QE tools and techniques to design and develop the new medical devices. Examples of such tools are Failure Mode and Effect Analysis (FMEA), Quality Function Deployment (QFD), Highly Accelerated Life Testing (HALT), requirements management (e.g., systems engineering), and SWOT analysis. In addition, many different tools and techniques have been used to improve these devices after they are released to the market. Most of these tools and techniques are predominately focused on showing compliance with Design Control or responding to field performance (e.g., Histograms, Pareto charts, Poisson trend analysis to analyze complaints and Medical Device Reports [MDRs]). This certainly raises the question: "Why not use the tools in an integrated way to not only comply with regulations, but also create distinct product capabilities and competencies?"

Up until a few years ago, it was typical for a design or manufacturing professional in the medical device industry to be told that their responsibility was to make sure that all current Good Manufacturing Practice (cGMP) guidelines were met by following written procedures. Many common-sense-based questions were never raised during product design and development. Some of these questions are: Who is to say that the procedures are correct? Where do the specifications come from? What is their relationship to customer requirements? Why can they not be changed? What specific product or technology standards were adopted and why? Why the specific interpretation of such standard? Did anybody have manufacturability in mind when they designed the device? How can special features incorporated in the design be measured? How are we going to make a profit in this business? These are some of the typical questions that a well-established Design for Six Sigma (DFSS) program can help answer by addressing them in a proactive mode as much as possible.

What is a medical device?

The definition of a "device" appears in section 201(h) of the Food Drug & Cosmetic Act. A device is:

"...an instrument, apparatus, implement, machine, contrivance, implant, *in vitro* reagent, or other similar or related article, including a component, part, or accessory, which is:

- Recognized in the official National Formulary, or the United States Pharmacopoeia, or any supplement to them,
- Intended for use in the diagnosis of disease or other conditions, or in the cure, mitigation, treatment, or prevention of disease, in man or other animals, or
- Intended to affect the structure or any function of the body of man or other animals, and which does not achieve any of its primary intended purposes through chemical action within or on the body of man or other animals and which is not dependent upon being metabolized for the achievement of any of its primary intended purposes ..."

The definition in ISO 13485 (2003) is any instrument, apparatus, implement, machine, appliance, implant, *in vitro* reagent or calibrator, software, material or other similar or related article, intended by the manufacturer to be used, alone or in combination, for human beings for one or more of the specific purpose(s) of:

- Diagnosis, prevention, monitoring, treatment, or alleviation of disease
- Diagnosis, monitoring, treatment, alleviation of, or compensation for an injury
- Investigation, replacement, modification, or support of the anatomy or of a physiological process
- Supporting or sustaining life
- Control of conception
- Disinfection of medical devices
- Providing information for medical purposes by means of *in vitro* examination of specimens derived from the human body
- Which does not achieve its primary intended action in or on the human body by pharmacological, immunological, or metabolic means, but which may be assisted in its function by such means.

If one goes by these definitions, it is obvious that the new breed of combination devices mentioned earlier may not be classified as medical devices. We believe that it is just a matter of time before the

FDA and the ISO either modify the current definition of devices or develop specific requirements for these new breed of devices.

The Design Control requirements (FDA regulation)

With the introduction of Design Control regulations by the FDA in 1997, all medical device manufacturers must comply with these Quality System Regulations if they want to sell products in the United States. Compliance with such regulations should provide appropriate answers to the questions above. On one hand, failure to comply can result in a medical device manufacturer being cited for noncompliance through "FDA 483s," warning letters, or other FDA enforcement actions. On the other hand, full compliance with these regulations can result in positive effects, including but not limited to:

1. Fewer customer complaints and MDRs
2. More satisfied customers
3. Faster time to market
4. Fewer manufacturing "deviations"
5. Fewer defects or scrap or rework
6. Less overhead in manufacturing operations and compliance groups

Needless to mention, these benefits can potentially lead to an increase in a medical device company's market share and profits. With the adoption of Design Controls, the medical device industry saw wider application of the tools of quality.* For example, the guidance documents from the Global Harmonization Task Force (GHTF; see www.GHTF.org) are among the few documents that use the tools of quality to address how to comply with Quality System Regulations. However, many of these tools are typically misused in part because there is no linkage to each other or to a common roadmap, since compliance with QSR is the predominant driver (e.g., to present quality system records and, yes, more paper). The following set of questions can help illustrate this point:

1. Do we know if the failure modes that are seen during design and development can be traced to initial customer requirements? Are these failure modes actual failures or were the user needs incompletely defined? Did anybody in the firm foresee the actual hazards?

* Also known as the tools of Six Sigma.

2. Do we know if the failure modes that are seen during design and development can be mitigated through proper process validation and control, or are these failure modes inherent in the design or the design requirements?
3. Are the parameters selected for optimization during process validation based on risk analysis? How do we know that no real risks are being ignored?
4. Is it really possible to predict field performance and reliability level prior to product release? Is it practical? What about complicated systems?

Since the tools are typically used without linkages, the answers to these questions are usually "no" or "maybe" or "nobody knows." In big companies, top management is not aware of these little important details, and middle management does not want to pass along the bad news. While this was happening in the medical device industry over the past few years, other industries were embracing roadmaps and methodologies to help them improve even more so that they could become world-class in their industry, if not all of industry. For example, Toyota uses lean manufacturing principles to reduce its inventory levels, and Dell is well known for its custom computer assembly operations (e.g., mass customization). If we look to a field performance database and analysis website such as www.consumer-report.org, we may notice the high level of quality and reliability that the products from both companies enjoy.

While achieving compliance with Design Control requirements is basic and paramount to all medical device companies, it is possible that they might be satisfied as long as they fully comply with such regulations. If this happens, there may be a short-term increase in market share or the product will merely be launched on time, but for sustained growth, device companies (small and large) must focus on innovation (e.g., product, process, and management), excellence beyond compliance, and continuous improvement.*

Does this mean medical device companies are too far behind other industries? Does this mean that medical device companies cannot become world-class in the near future? Another disheartening fact is that since the start of the Malcolm Baldrige National Award in 1988 in the United States, very few medical device or pharmaceutical companies have ever won it (see www.qual-ity.nist.gov/Contacts_Profiles.htm). Lastly, it is said that the largest

---

* As explained later in Chapter 4, in the Six Sigma world it is recognized that you can improve quality one project at a time, thus, continuous improvement implies a portfolio of projects.

component of the cost of goods sold (or product cost) and company overhead is Quality Assurance (QA).

How do medical device companies manage fierce competition, be compliant with the regulations, and release good quality products to the market at the same time? How do medical device companies "catch up" to other world-class companies and yet maintain the innovation and flexibility that have helped them grow faster compared to other industries? How do medical device companies use superior performance, engineering and scientific knowledge, and reliability as obstacles to competition?

The answer to these questions may very well be "Six Sigma."

Six Sigma and design for Six Sigma: what is it?

In the recent past, there has been tremendous focus on Six Sigma initiatives by many different companies in various industries. While there are many definitions for Six Sigma, the technical definition for it can be "a structured approach to improving a product or process to result in only 3.4 defects per million opportunities." Another simple definition is the quality of the business.

It must be mentioned that ever since big corporations such as General Electric and Motorola have embraced this initiative, it has moved beyond just being a quality improvement initiative. It is one of the few technical initiatives that have caught the attention of business leaders. The book *Six Sigma, The Breakthrough Management Strategy* by Harry and Schroeder became a *New York Times* bestseller and, in fact, can be found in the business shelves in airport bookstores.

Six Sigma is now being treated as a philosophy, modern management system, or "way of life" of an organization that wants to be seen as a source of value creation and wealth. When we say way of life, we mean that some companies use Six Sigma philosophies to run their day-to-day operations as well as the roadmap to achieve strategic objectives. For example, Becton & Dickinson, a New Jersey-based medical device company, announced the following in its 2002 annual report to shareholders: "Our Six Sigma quality program has completed its second year with more than 170 'Black Belt' experts and an active 'Green Belt' training program."

What is DMAIC and why is it said to be reactive?

DMAIC stands for Define, Measure, Analyze, Improve, and Control. The DMAIC methodology is typically used for improving existing products and processes in a company. Specifically it is used when low yields, high scrap, or simply poor customer satisfaction indicate potential problems in the execution of the manufacturing steps or service provided by a company.

The DMAIC methodology is almost universally recognized and defined as comprising of the following five phases: Define, Measure, Analyze, Improve, and Control. In some businesses, only four phases (Measure, Analyze, Improve, and Control) are used; in this case, the Define deliverables are then considered prework for the project or are included within the Measure phase. The DMAIC methodology breaks down as follows:

- Define the project goals and customer (internal and external) requirements.
- Measure the process to determine current performance.
- Analyze and determine the root cause(s) of the defects.
- Improve the process by eliminating defect root causes.
- Control future process performance.

While there are certainly gains made by many medical device companies, we strongly believe that in order to grow their business, these companies must properly apply proactive methodologies such as Design for Six Sigma. Many books have been published so far that explain both the technical and business aspects of it. These books focus mostly on applying this initiative for manufacturing or transactional processes. Only a few of them focus on applying Six Sigma to design and develop products and its processes. In any event, to our knowledge, there are no books that specifically focus on applying Six Sigma to medical device design and development.

We have observed, applied, and championed both Design Control and Six Sigma concepts. Given the nature of the medical device industry, it is not surprising that medical device companies struggle with the idea of implementing initiatives such as Six Sigma to product design and process improvement, especially after the products are approved for sale. This can be even more prevalent in companies that have devices that must go through FDA's Pre-Market Approval (PMA) process.

By writing this book, we want to fill the void in the availability of published material in application of Six Sigma for medical device design and development. We provide a meaningful linkage with

FDA's Design Control guidelines to companies that have to be compliant with regulations, including Design Controls. As a result, a medical device company could not only adopt the Six Sigma philosophies and tools but can also be on the right track to comply with the Quality System Regulations. We also provide sufficient clarity on the design, development, validation, and control of the manufacturing processes that make devices.

In Chapter 2, we briefly focus on FDA's Design Control roadmap and its implementation. For a detailed look at this topic, we encourage the readers to refer to our book, *Practical Design Control Implementation for Medical Devices*. In Chapter 3, we focus on the Six Sigma roadmap for product and process development. Quality Engineering tools and their linkages to the Six Sigma roadmap are introduced. After explaining both these concepts separately, we show in Chapter 4 how both Design Control and Six Sigma roadmaps can be linked for maximum effectiveness. In Chapter 5, we provide details on pitfalls to avoid in implementing both these roadmaps. We strongly urge readers to pay special attention to the contents of this chapter, since it highlights certain beliefs and behaviors unique to medical device companies. These beliefs can reduce the effectiveness of Design Control and Six Sigma roadmaps. Implementation of these roadmaps calls for a means to measure the effectiveness of these roadmaps. This is our focus in Chapter 6.

Finally, the book's appendices provide sample Design Control and Six Sigma plans for product and process development (commercial and clinical release).

The primary audience for this book is anyone responsible for developing and implementing a product or process to comply with FDA's Design Control regulations. This includes engineers in product and process design, development, and implementation in small, medium, and large medical device companies in the United States as well as those outside of the United States that sell products in this country. This book can also be used by companies that have implemented or are in the process of implementing a quality system for Design Control. The book also serves the needs of other product or process development team members including, but not limited to, representatives from marketing, quality, regulatory compliance, and clinicals.

Diference between Six Sigma programs and the regulations

Regulations exist for the purpose of protecting the people, not to boost the financial wealth of anybody. On the other hand, the main reason for adopting a Six Sigma program has been purely of financial benefit. Though regulations and Six Sigma seem to be the antithesis of each other, the reality is that well-executed product development projects can satisfy both. In all our years of experience in the medical device industry and after reading warning letters issued to companies as well as 483 reports, we infer that the main underlying reason for most of the observations made on large companies has its roots in lack of knowledge and understanding about the products they make and the technologies they use. This lack of knowledge is exacerbated by two paradigms:

First, the false belief that implementing an adequate quality system (e.g., policies, procedures, organizational structure, accountability) will ensure safety and effectiveness of the medical devices being designed or made. For example, it is amazing to watch the amount of money invested by medical device manufacturers developing Corrective and Preventive Action (CAPA) programs and systems and to observe how these systems are typically applied. Let us provide you with a scenario that will highlight the false belief with respect to CAPA. Our conversations with many medical device professionals at conferences show that technical and scientific knowledge of the product is typically not present within the reach of the factory. This may explain why the typical root cause mentioned in many CAPA reports is "operator error" and the typical corrective action mentioned is "retrain operator."

This may also explain why many CAPA systems contain complaints with reports stating, "Problem could not be reproduced." We think these are typical signals of lack of true understanding and knowledge of the product and the belief that just having a CAPA system can protect the product and the company over time. Medical device companies have to face the reality of today's job markets: It is difficult to find experienced and knowledgeable technical professionals who passionately know the product, its design, its manufacturing process, and its applications. Who can then answer the question, "What is wrong with the product if it meets the specifications?" with full authority?

Second, the false belief that suppliers or contract manufacturers know what they are doing. The buzzword "supply chain" was made famous by the 1993 Harvard Business School case study *Liz Claiborne, Inc. and Ruentex Industries, Ltd.** This famous case study inspires business leaders and brand-new MBAs to believe that all can be "contract manufactured" by somebody with extra capacity. Nowadays, it is not uncommon to see medical device companies hire contract design and development houses that have the capacity to design and develop. The fact remains that, very simply, extra capacity means you are not the "bread and butter." One might contend that these contract manufacturers also provide sufficient documentation to meet QSR requirements. We are certainly not against utilizing contract houses, but we just wanted to highlight the second false belief. Will these contract design and manufacturing facilities understand and care about your product as much as you do? How are you ensuring that this happens throughout the product life cycle and not just the product development life cycle?

A well-conceived and -implemented Six Sigma program will evaluate all business paradigms and fallacies and will objectively reveal the sometimes-painful truth of unreal management optimism. It will find alternative solutions, show its business case, get it implemented, and move to the next opportunity.

---

* Harvard Business School case study # 9-693-098, 1993 (abridged).

chapter two

Design Control roadmap

This chapter is aimed at introducing Design Control requirements while attempting to show that good understanding of such regulations is no insurance to designing safe and effective medical devices. Design Control requirements, part of the Food and Drug Administration's (FDA's) Quality System Regulations (QSRs), went into effect on June 1, 1998. Before this period, medical device companies selling their products in Europe had been required to comply with the Design Control requirements of ISO 9001and the EN 46001 standards. Design Control is one of the four major subsystems in the Quality System Regulations.*

What is Design Control?

Design Control can be seen as a set of requirements, practices, and procedures incorporated into the design and development process and associated manufacturing processes for medical devices to ensure that they meet customer, technical, and regulatory expectations. In our first book, *Practical Design Control Implementation for Medical Devices*, we added "disciplines" to this definition. Table 2.1 depicts the Design Control requirements and typical associated quality systems (Gopalaswamy and Justiniano). These quality systems can be seen as one element of the firm's mechanism to comply with the regulation. Simply put, Design Control helps a medical device company understand regulatory compliance requirements (the "whats"

---

* The other three are Management Controls, Corrective and Preventive Actions, and Production and Process Controls (see QSIT guide in www.FDA.gov).

of what the customer wants or needs*) and create a quality system to meet those requirements. Notice that knowing the regulation is not a guarantee of developing safe and effective medical devices. There is still the need to know the life science facts (e.g., anatomy, biochemistry) and to have the scientific or engineering capabilities and resources to be able to adopt existing technologies** that will turn into new medical applications.

What is not Design Control?

Design Control is certainly not a detailed step-by-step prescription for the design and development of medical devices. It does not help a company, rightfully so, by providing tools and methodologies to consistently meet regulatory and customer expectations (the "hows").

Furthermore, it does not challenge the science, the development "modus operandi," the "inventive stages," or the engineering knowledge in product design and development. FDA investigators will evaluate the process, the methods, and the procedures for Design Control that a manufacturer has established.*** The regulation allows for extensive flexibility in the systems for Design Control due to the wide variety of medical devices and the technologies involved.

Implementation of a Design Control system has difficulties. First, there is a strong possibility that the design and development organization will oppose putting in place more than minimum rigor. This is especially true when the compliance team is composed of people with little or no knowledge at all about the design, the clinical procedures, or the technologies involved. The design and development team may feel overwhelmed with "people who do not understand looking over their shoulders and requesting paperwork." In Chapters 3 through 5, we will bring up DFSS tools such as the design cascade that will facilitate the design communication between the team and other functional organizations or auditors.

Even if they agree and are a willing organization, the Design Control implementation team must follow detailed steps. Some of the necessary steps, in chronological order, are:

* In its role of protecting "The People," the FDA's underlying "what" is safe and effective. DFSS will help companies to identify the technical features needed to achieve safety and effectiveness.
** Throughout this book we will emphasize the fact that most medical devices are based on existing technologies. For example, most plastics and metal alloys already exist for commercial purposes. The innovation brought in by the medical device design company is merely the specific applications to a human medical need.
*** See the 1999 "Guide to Inspections of Quality Systems" (the QSIT manual) in the CDRH section of FDA's website, www.FDA.gov.

1. Define the design and development process of the firm. For example, technology development and discovery, concept development and feasibility, design works, prototyping, testing, pilot runs, reviews, etc. The firm shall define where design controls will really start applying and also when a product is formally released to manufacturing (design transfer). In Chapter 4 the DFSS methodology will be related to the design and development process. Very clearly state when design control does start.
2. Development of policies, procedures (see Table 2.2), and work instructions for appropriate control of the design and development process of the device and its manufacturing process.
3. Development of policies, procedures, and work instructions for risk analysis.*
4. Development of training plan. Typical skills where medical device companies need to strengthen are quality systems for "non-quality personnel," compliance with the regulation, reliability engineering, use of external standards, Six Sigma methodologies (e.g., DFSS), FMEA, FTA, and statistical methods for non-statisticians.
5. Definition of internal and external interfaces and roles. For example, if a new manufacturing process has to be developed, then the product development team will need a manufacturing process development engineer or specialist interfacing with it. Of practical importance is the determination of who those interfaces are in terms of scientific, technical, and medical or clinical expertise.
6. Review quality systems for adequacy. For example, a company that has been manufacturing plastic and metal-based devices is moving faster towards designing capital equipment with electronic components.

Note that Six Sigma is mentioned as a typical missing element in the medical device industry.

Design Controls and IDE

Originally, devices being evaluated under Investigational Device Exemption (IDE) were exempted from the original Good Manufacturing

---

* The regulation states risk analysis; however, further clarifications from FDA clarified that the actual requirement is risk management (refer to ISO 14971).

Table 2.1 Design Control requirements (21 CFR Part 820.30) and typical associated quality systems

Requirement	Typical associated quality systems
a) General	Preparation of quality policies and procedures related to Design Controls and other associated quality systems.
b) Design and development planning	Specific procedures[a] for product design and development planning and design change planning.
c) Design input	Procedures[b] for data collection, analysis, and storage (filing) on customers, users, installers, historical field quality data. For example, focus groups with doctors or other healthcare givers, panel discussions, interviews, surveys, field complaints, MDRs, human factors engineering (e.g., ergonomics, industrial design). How to execute, document, analyze, and store such information. A key procedure is the one that indicates how design inputs are approved (ideally stated in the design and development plan).
d) Design output	Procedures for translating design input into engineering or scientific design specifications. How to execute, document, analyze, and store such information. Procedures for planning, executing, and documenting experimental protocols such as design verification and validation. A key procedure is the one that indicates how design outputs are approved (ideally stated in the design and development plan and according to verification and validation of design).
e) Design reviews	Procedures for organizing, executing, and documenting design reviews. Procedures for defining a design and development team roster and their reviewers. How to document pending issues and how to follow up and close all of them. How to execute, document, analyze, and store such information.
f) Design verification	Procedures for software/hardware verification. How to execute, document, analyze, and store such information.

(continued)

Table 2.1 Design Control requirements (21 CFR Part 820.30) and typical associated quality systems (continued)

Requirement	Typical associated quality systems
g) Design validation	Procedures for software/hardware validation, animal studies, clinical studies, cadaver laboratories. How to execute, document, analyze, and store such information.
h) Design transfer	Procedures for the preparation of DMR, process validation (IQ/OQ/PQ), training. Supplier or contract manufacturer certification. How to execute, document, analyze, and store such information.
i) Design changes	Procedures for changes and updates to "pre-production." How to execute, document, analyze, and store such information.
j) Design history file	Procedures for creating, approving, and updating the DHF. How to execute, document, analyze, and store such information.

[a] The regulation does not ask for a design plan procedure. However, in practical application it is a good product design practice to have procedures or guides that define how a firm designs, develops, and controls the design requirements. In DFSS we will talk about methodologies such as CTQ cascade, QFD, and IDDOV (the hows). AdvaMed (May 15, 2003) states that most companies incorporate all the requirements for each of the elements in the first column into one overall design and development procedure.

[b] The procedures are useful to standardize creation of the elements of the DHF. The methods for gathering and analyzing customer inputs such as focus groups, surveys, and conventions are purely dependent on the skills of those doing the work. Here is where many of the tools of DFSS become useful as discussed in the next chapters.

Process (GMP) regulation. This makes sense for two reasons: such devices were mostly "lab made," with really no production equipment or no GMP mass manufacturing processes applied at that time (even though sponsors were required to ensure manufacturing process control); also, the devices might never be approved for commercial distribution.

Our past lives in the electronics, automotive, and telecommunications industries tell us that even for investigation purposes, it is a sound strategy to develop and build experimental models or prototypes under Design Controls. It is also the idea from FDA. In an IDE evaluation, human beings are being exposed to the potential hazards of an IDE device. Also, IDE data may be used as evidence of design verification or design validation. Results from IDE usually motivate design changes. Applying the disciplines of Design Control can

Table 2.2 Design input examples

Potential design input	Examples
Intended use	Specific vs. general surgery instrumentation. Endoscopic or open surgery? Screening or final abused drug immunoassay. Beating or still-heart surgery.
User(s)	Installer, maintenance technician, trainer, nurse, physician, clinician, patient. What is the current familiarity of all potential users with this technology?[a]
Performance requirements	Highly sensitive immunoassay or with a very broad dynamic range. How long will the surgical procedure last?[b] Is there a potential complication with very big, very small, obese, skinny, very ill, very old, very young patients? Frequency of calibration longer than a month for a diagnostic assay. Is it part of a typical battery of assays (e.g., T4, T3, and TSH)? Software user interface requirements. Software requirement specifications.
Chemical/environmental characteristics	Biodegradable packaging.
Compatibility with user(s)	Biocompatibility and toxicity.
Sterility	Pyrogen-free. Sterile. Is it to be used within the sterile field in surgery?
Compatibility with accessories/ancillary equipment	IV bag spike or other standard connectors. Electrical power (e.g., U.S. vs. South America). Open architecture for computer systems networking. Is it to be used with an endoscope? Which channel size(s)?
Labeling/packaging	Languages, special conditions, special warnings (e.g., C 60601-1 states several warning symbols). Heat protection. Vibration protection. Fragility level.
Shipping and storage conditions	Bulk shipments or final package. Humidity and temperature ranges.
Ergonomics[c] and human factors	International vs. domestic considerations. "Fool proof."

(continued)

Table 2.2 Design input examples (continued)

Potential design input	Examples
Physical facilities dimensions	Power cables for electrosurgical generators may need to be longer in Europe than in U.S. operating rooms.
	The same would apply to devices that include tubing for blowing CO_2 (e.g., a blower with mist for CABG that is used to clean arteriotomy area from blood). The length of the tubing had to be longer for Europe than for U.S. operating rooms.
Device disposition	Disposable vs. reusable.
Safety requirements	UL/IEC/AAMI and country-specific requirements.
Electromagnetic compatibility and other electrical considerations	Electrostatic discharge (ESD) protection.
	Surge protection.
	EMI/EMC (meet IEC standards) for immunity or susceptibility.
Limits and tolerances	Maximum allowable leakage current on an electronic device.
Potential hazards to mitigate (risk analysis and assessment)	Potential misuses such as warnings or contraindications in inserts or user manuals.
	Hazards in absence of a device failure (e.g., electrocution of an infant with metallic probes of a device).
Compatibility with the environment of intended use	A metallic surgical device that may contact an energy-based device during surgery could conduct energy, thus potentially harming the other organs of the patient.
Reliability requirements	99% reliability at 95% confidence at the maximum usage time or conditions.
	Mean time between failures.
	Mean time to failure.
	Mean time to repair.
	Ease of self-repair and maintainability.
	Mean time to maintenance.
User(s) required training	Simplify new surgical instrument and new procedure because it may require complicated training.
	Programming a hand-held blood sugar analyzer. How intuitive is a new table computer software for clinical data entry? How does the user know that the data has been saved to the server?
MDRs/complaints/failures and CAPA records and other historical data	Benchmark from similar, platform, or surrogate device. Use the MAUDE database from www.fda.gov to search for MDRs and reported adverse events.[d]

(continued)

Table 2.2 Design input examples (continued)

Potential design input	Examples
Statutory and regulatory requirements	Policy 65 (California).
Physical characteristics	Dark color in an endosurgical or laparoscopic instrument to avoid reflection of light from endoscope.
	Amber- or dark-colored bottles for filling of light-sensitive reagents.
Voluntary standards	IEEE for electrical components or software development and validation.
	NCCLS for IVD.
Manufacturing processes	Design the device such that no new capital equipment is required for manufacturing.

a This design input can directly identify a design output such as training requirements. By the way, not only training for users, but businesswise for sales and marketing personnel.

b This is especially important to define the use environment that eventually defines the required reliability. This is an example of questions that the R&D quality or R&D reliability engineer should be asking during the gathering of design inputs.

c Consider all users. For example, you gather data among males, ignoring female users, for a device that may require some sort of hand activation.

d For example, in a 510(k) device, it would be of no value added not to consider the typical malfunctions and performance or safety issues of a predicate device.

ensure controlled design changes and better evaluation of results. Therefore, FDA has amended the IDE regulation (CFR 812), and thus Design Controls do apply to IDE devices.

Design Control requirements

This section introduces and discusses each Design Control requirement. Practical examples are provided as well as a table that relates each 21 CFR 820 requirement with ISO 9000. Typical quality systems audit questions are provided with the purpose of helping the firms to execute their own self-assessment.

Design planning

Plans shall be produced that allocate the responsibility for each design and development activity. Somehow, the design and development process has to be controlled, and the product development team or R&D management shall have a sense of where they are with respect to project design goals and time. Each of these activities shall be referenced and described within the plan. This shall be an ongoing process until the design is completed, verified, and validated.

Whoever is in charge of generating the design and development plan (DADP) shall keep in mind that the underlying purpose is to control the design process aimed at meeting the device's intended use and its associated quality objectives.

Benefits associated with a design and development plan

1. Disciplined approach to project management. Thus, knowledge-based decision making becomes plausible. DFSS tools and philosophy will help to make this very handy.
2. Project specific (e.g., specific details).
3. Common communication mechanism (e.g., "everybody is on the same page").
4. Proactive planning (e.g., no surprises to the interfaces or top management).
5. Regulatory, marketing, economic (e.g., cost of manufacturing), and quality requirements are included in one structure that facilitates alignment for all parties involved or with a stake in the project. This is the chance to bring the organization together and adopt the new terminology (e.g., Device Master Record [DMR], Design History File [DHF], design validation).
6. Ease for project issue resolution.
7. Overall compliance record and traceability (e.g., Why did we do it like that?, Why did we choose material A over material B?).

Organizational and technical interfaces

Several groups of personnel may provide input to the design process, and it is essential that any organizational and technical interfaces between these groups are clearly defined and documented. For some firms, these interfaces may have a role of ownership, as technical leadership, or as subject matter experts for some of the project's milestones or phases. This information should be reviewed regularly and made available to all groups concerned. Technical interfaces are an interdependent part of 820.30 b – Design and Development Planning.

Design input

The purpose of all products should be clearly understood so that the design inputs can be identified and documented. The company shall review these inputs, and any inquiries should be resolved with those

Table 2.3 Examples of how to go from raw design input into design requirements (design requirements cascade)

Practical interpretation →	Customer requirements External customer needs and internal goals	System requirements Measurable customer requirements	Design input Design requirements[a]
Example 1:	... can be used on big and small humans.	Targeted at 90% of domestic potential patients	Standard small, medium, and large sizes based
Example 2:	... most reliable device in its class.	Total reliability = 99.7%	Reliability allocation for three main subsystems = 99.9%[b]

[a] Note that at this level of the cascade of requirements, we say it is a design requirement, not the engineering specification yet (design output).

[b] Using simple principles of probability in systems reliability, the three main subsystems are allocated the same reliability of 99.9%, thus .999 x .999 x .999 = .99.7%.

responsible for the original specification. The results of contract reviews should be considered.

Design input can be defined as performance, safety, business economics, and regulatory requirements* that are used as a basis for device design. From Table 2.2, we may realize that design input comes in many ways. The two examples in Table 2.3 give us an idea that when we hear the customer, we are not going to get "direct usable" design inputs. Such information has to be interpreted and massaged (e.g., in DFSS terminology it is "cascaded") to be able to specify design requirements. For example, you can never expect a customer to tell you what kind of plastic resin you have to use to meet some need for a medical device. In the language of DFSS, the most important inputs will be called Critical to Quality (CTQ).

Design output

The objectives of any new product design should be defined as design outputs. These should be clearly understood and documented. They should be quantified and defined and expressed in terms of analyses and characteristics. A very important requirement is traceability to

---

* For some devices, there are specific requirements stated in medical device standards such as IEC 60601-1.

Table 2.4 Example of design output meeting design input (cascade)

Design input	Design output	
	Design specification	DMR
The medical device will be used in trauma rooms. It must be capable of withstanding adverse conditions (e.g., accidental pulling by the tubing).	The bond strength between a luer lock and tubing (IV line) shall withstand P pounds of axial force without detaching from the tubing	The raw material for the luer lock will be X and the solvent Y. Before inserting the tubing into the luer lock, the solvent will be applied and a curing of T minutes will be allowed.

design inputs. It is here where the DFSS CTQ cascade can become a regulatory compliance deliverable (see Table 2.4).

Examples of design output

1. The device
2. Labeling for the device, its accessories, and shipping container(s)
3. Insert, user manual, or service manual*
4. Testing specifications and drawings (detailed, measurable)
5. Manufacturing (materials and production) and QA specifications or acceptance criteria
6. Specific procedures (e.g., manufacturing equipment installation, work instructions, BOM, sterilization procedures)
7. Packaging feasibility studies, validation testing, and results
8. Risk analysis, risk assessment, FMEAs, reliability planning, and results
9. Biocompatibility and toxicity results
10. Software source code
11. Software hazard analysis
12. Software architecture
13. Software Verification and Validation (V&V)
14. The 510(k), IDE, or PMA submission
15. Technical file or design dossier for Clinical Evaluation (CE) marking
16. Clinical evaluation results

* Service manual usually contains instructions for repairs and preventive maintenance. Mainly applies to capital equipment such as MRI, CT scan, electrosurgical generators, diagnostic analyzers, etc.

17. Transit, storage, and shipping conditions testing and results
18. Supplier and component qualification (e.g., the DHF shall include evidence of official communication to component suppliers stating status of qualification approval and process control agreements*)

In general, the design output deliverables will reside in the DHF and the DMR.

What is the relationship among design input, design output, DHF, and DMR?

Figure 2.1 shows that design output is really an answer to a request (design input) plus the evidence to support the decision. From the list above, we can say that all those design outputs belong in the DHF at any given time, as depicted in Figure 2.2. However, only items 1 to 6 would end up being part of the DMR. The DHF can be seen as a "virtual file," with records showing the relationship between design input and design output. The key word is "records." The DMR is composed of the instructions and criteria needed to make the product. While the DHF is made of records, the DMR is made of "living documents."

Design review

Competent personnel representing functions that are concerned with the particular design stage under review conduct design reviews at various (sometimes predetermined) stages of the design and development process. There are two key elements in design review. The first one is the independence between reviewers and design and development team members. This in fact is the principle behind quality audits and assessment. Independent eyes and ears are not "biased." The second one is the value that a given reviewer brings to the design and development project (e.g., technical, medical, and clinical knowledge, among others). Quality system procedures must ensure that these reviews are formal, planned, documented, and maintained for future review. If a firm adopts DFSS, the commonly known tools could facilitate design reviews. For example, a typical first design review is mostly aimed at assessing the Voice of the

---

* This is another example of the need for appropriate quality systems. There should be a procedure to evaluate and qualify suppliers. It shall also describe what documentation is required to notify the supplier when qualification has been attained.

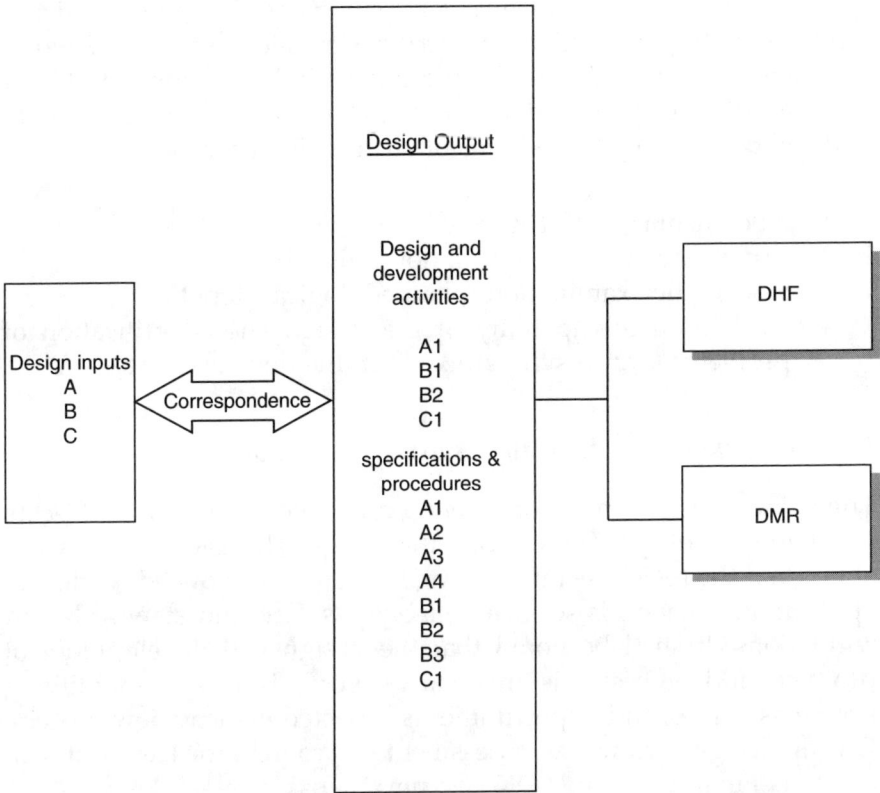

Figure 2.1 Relationship among design input, design output, DHF, and DMR.

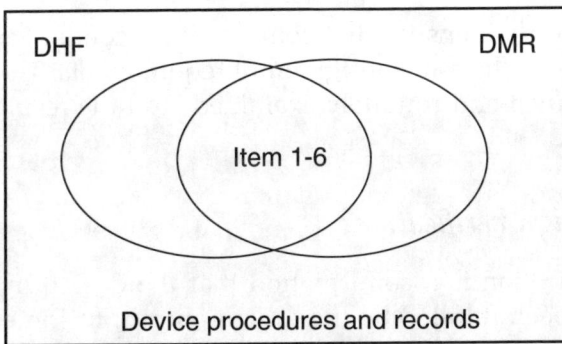

Figure 2.2 Some DHF elements will become DMR elements.

Customer (VOC). This is the main design input to any given project. Records must be kept as part of the DHF. The requirements cascade or traceability matrix can be used to keep track of reviewed design items.

Formal documented reviews of the design results are planned and conducted at appropriate stages of the design and development work. Such stages are to be defined by the design and development plan or the design change plan.* Reviewers shall have no direct responsibility (independence). Key ingredients of such reviews are:

1. Documentation (formal)
2. Comprehension (technical, not political)
3. Systematic examination (planned, logical steps)
4. Evaluation of capability of the design and identification of problems (not to sympathize with the development team)

Design review: practical needs and value added

The value added comes from having an independent body of peers ("different set of eyes") reviewing the design. This is especially valuable when the review team is strong in customer knowledge, clinical applications, materials science, reliability, safety, and standards and regulations. It shall be noted that the design and development of products and processes is an iterative work. Therefore, identifying problems, issues, and opportunities is expected in the review process. During design reviews, an assessment of progress (or lack of it) can be done. Finally, it is the "OK" for next steps.

Design verification

The company shall ensure that competent personnel verify that the design outputs satisfy the design input requirements. These activities must be planned and routinely examined, and the results should be documented.

What is design verification?

Design verification is a confirmation that the design input requirements have been fulfilled by the design output. In some companies, common sense drove them to adopt similar concepts such as "Design Engineering Pilot," "Design Pilot," and "Engineering Built." The regulation aims at providing a sense for formality (i.e., procedures) and

---

* Thus, design reviews apply to product already in the market that is being exposed to a design change.

Table 2.5 Example of design input, output, and verification

Design input	Design output		Design verification
	Design specification	DMR	
The medical device will be used in trauma rooms. It must be capable of withstanding adverse conditions (e.g., accidental pulling by the tubing).	The bond strength between a luer lock and tubing (IV line) shall withstand P pounds of axial force without detaching from the tubing.	The raw material for the luer lock will be X and the solvent Y. Before inserting the tubing into the luer lock, the solvent will be applied and a curing of T minutes will be allowed.	At 99% reliability and 95% confidence, a Safety Factor of 3 was obtained during a stress-strength test.

structure (i.e., design plan*) within the DHF. As indicated in Table 2.5, the firm should prepare itself for successful design verification by defining quantifiable design inputs and their corresponding design outputs. In our first book we devoted a section showing how a design history matrix can help manufacturers to plan and execute an acceptable design verification. In the DFSS world, such a table is called a requirements cascade or a requirements traceability matrix. This tool not only will help the firm to comply with the regulation, but it is also a good design engineering practice.

Design validation

To ensure that the product conforms to customer requirements and defined user needs, it is essential that design validation be undertaken. This will normally be performed on the finished product under defined conditions. If the product has more than one use, multiple design validations may be necessary.

Design validation always follows successful design verification. Design verification is done while the design work is being performed. The medical device may not be complete or may not be in its final configuration. To validate design, the team needs to have the final medical device.

* Beyond the generality of the design plan, there shall be performance, quality, and reliability goals already established. Some firms may decide to include all the project requirements in the design and development plan; others may decide to establish interdependent quality and reliability plans in addition to the design and development plan. The same would apply to the design change plan.

Design validation includes software and the hardware–software interface by challenging the source code in its actual use conditions. For example, embedded software verification is done by emulation of the source code, while software validation is done once the software has been "burned" into the chip or EPROM and the system is challenged.

Why design validation?

If verification demonstrated that design outputs met design inputs, why are we validating design? This question can be answered simply by saying that the design inputs may not be the real thing. That is, they do not lead the design to meet the customer needs. Also, even if design inputs are right, then the design outputs could be wrong. One possible reason could be changes in customer requirements since the design was initiated. If design inputs and outputs are right, then there could had been a problem when the design was transferred to manufacturing.*

Notice that design validation is a final challenge to all the existing quality systems including Design Control, training of manufacturing personnel, process validation, and so on. In fact, design validation is the link to process validation. Also notice that if on one hand, the output from the process does not meet customer needs and intended use, the manufacturing process becomes worthless. On the other hand, if the process is not repeatable or reproducible, then the design validation is also worthless, since there is no manufacturing process that can ensure equivalent performance from unit to unit or from batch to batch.

Design transfer

Design transfer according to the regulation

"Each manufacturer shall establish and maintain procedures to ensure that the device design is correctly translated into production specifications" (21 CFR Part 820.30.h). The key words here are "correctly translated." This means that an auditor with technical knowledge of the product and process could find a connection between design outputs and what is stated in the DMR. For example, while working with external manufacturers, one of the authors realized that the ranges used in worst-case analysis were different from those

---

* This is the main reason why initial production units are the best choice for design validation.

stated in the drafted DMR. The reason was an opportunity for a cost reduction. The new ranges would imply an extrapolation; that is, beyond the worst-case analysis range examined for the Operational Qualification (OQ). Therefore, the change order to release the DMR had to be rejected and the right ranges were put back.

Practical definition of design transfer

What is really being transferred is knowledge from the design and development team to the manufacturing or process validation team. It is of utmost importance that the process validation team (e.g., those responsible for the "mass production") understands the device and its intended use as the first step. DFSS will help to transfer knowledge via the requirements cascade. It is also relevant the amount and kind of knowledge that the design and development team have about the manufacturing process. Careful attention shall be paid to what is done by R&D and what is to be done by manufacturing development or process validation personnel.

For example, we can think of process development in two stages. First, let us consider the design of a new manufacturing machine or piece of equipment. The design of the new machine is typically done by a design and development team, sometimes in combination with a consultant or with the machine manufacturer. But designing a manufacturing machine is not synonymous with process characterization. That is, there is usually a lot of unknown behavior from the newly designed machine. It is through Design of Experiments (DOE) and other DFSS tools used during process characterization that technical manufacturing personnel could really learn what the input parameters are that affect the output characteristics and in which way. In other words, if the manufacturing personnel know what the input parameters are (e.g., independent variables) and how they affect the output characteristics (e.g., dependent variables), then they may have a way to control the manufacturing process.

Summarizing this section, when new manufacturing equipment is designed, there are two main development steps involved; first, the design per se of the equipment. Second, characterizing the equipment or machine. Typically, the second step is a function of the "pilot plant." If a company is operating in a "direct to site" mode, then the second development step will have to take place at the manufacturer's site. This second development step falls under the definition of OQ, specifically under the concept of process characterization.

Design transfer may occur via documentation, training, R&D personnel sent to manufacturing, or manufacturing personnel having been part of the design and development team. All the design transfer activities shall be listed in the design and development plan. However, training and documentation do not fulfill the whole purpose of design transfer. The expected results of effective design transfer are:

- Product manufacturability and testability.
- Process repeatability (item to item, within a batch or lot).
- Process reproducibility (lot to lot).
- The process is under statistical control (stable), and thus it is predictable.
- Manufacturing personnel know what they are doing and what process parameters need to be adjusted, how to adjust them, why, and when (here, DFSS will help).
- Adequacy of DMR documents.
- The manufacturing and acceptance specifications are realistic and meaningful.
- Raw materials and components perform as expected.
- Suppliers know what they are doing.
- There are no surprises.
- A manufacturing process that consistently ensures a medical device that is safe and effective.

Design changes

All design changes must be authorized by people responsible to ensure the quality of the product. Procedures shall be established for identification, documentation, and review of all design changes. These must follow the same rigorous procedure adopted for the original design.

There are four elements involved in controlling design changes. The matrix in Table 2.6 depicts them.

Document control is a straightforward, classical GMP quality system for existing products. It is aimed at enumeration, identification, status and revision history of manufacturing specifications, testing instructions, and BOM (i.e., all elements of the DMR). However, once the design controls are part of the quality systems, the firm needs to control the documentation that is being "drafted" during design and development. Many temporary documents exist during the stages of product design; most of them will be subject to multiple changes. Thus, the big question is: How can the design and development team ensure harmonization between the already-approved design

Table 2.6 Design changes and the product life

	During design and development	After product has been released to the market (existing products)
Document control		
Change control		

elements (e.g., design reviews) and new elements of change while still going through design iterations or doing design verification or validation ? Figure 2.3 depicts design changes with a diagonal line that implies multiple changes in this temporary or conditional DMR during the entire design and development life cycle of the device. It is important to realize that not only DMR, but also elements of design verification and validation, can be affected (and thus, the DHF).

Change control per se has to do with the physical characteristics of the device, or its acceptance criteria or its testing or evaluation methods. For product under development, there has to be a logical procedure to expose the entire design and development team as well as reviewers to the changes. This is very much in line with the last two statements of the previous paragraph.

A bigger challenge in terms of regulatory compliance and business risk is the control of design changes on existing products. The changes can not only alter the design, but also the intended use (and thus the 510(k) or PMA submission to FDA). Another possibility is the change affecting some other device or subsystem manufactured. Our greatest concern in this situation is the fact that manufacturing operations are typically the ones requesting the changes in response to raw material or component deviations. Without competent personnel with access and understanding of the DHF, how can approvers of change be able to make a conscious decision? Also, manufacturing operations may never have the means for executing a design "re-validation" upon design changes. In this book we will introduce the DFSS concept called design requirements cascade, which is in line with classical 1980s system engineering programs.* Later, in Chapter 6, we will talk about the abuse of the design requirements cascade and other DFSS tools.

---

* Thus, nothing new about the concept or tool.

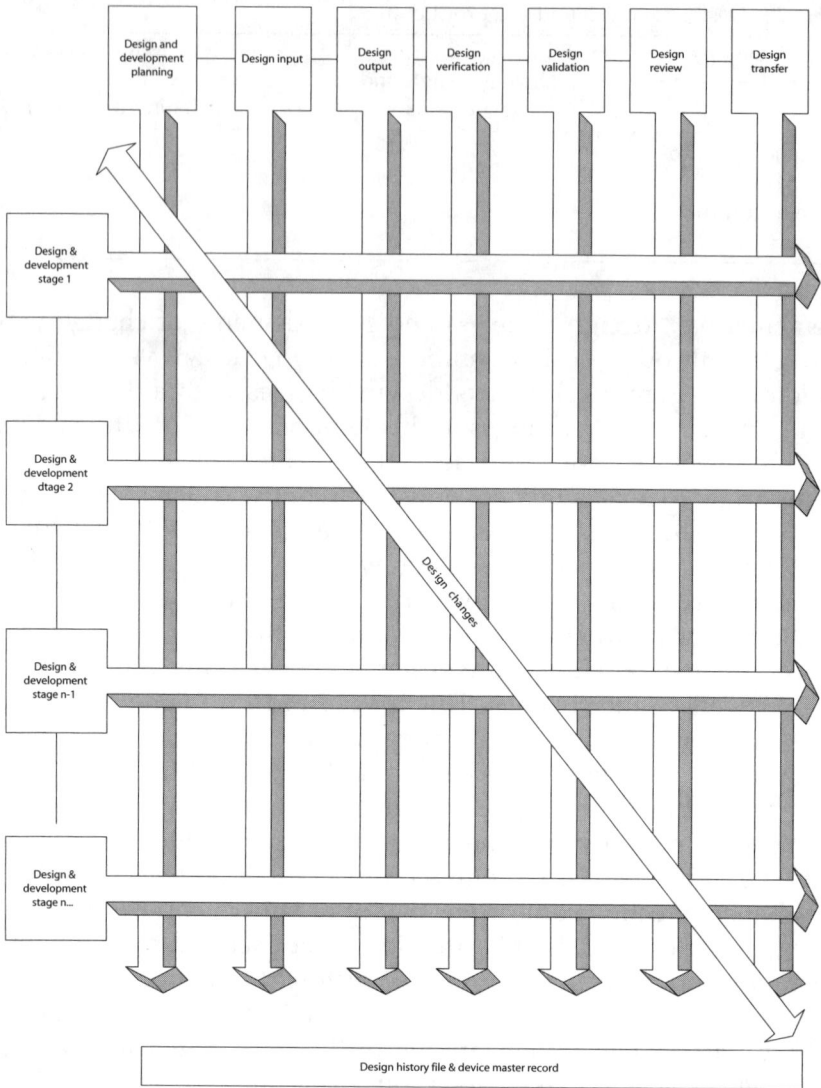

Figure 2.3 Design changes during design and development stages.

References

AdvaMed, May 15, 2003, "Points to Consider when Preparing for an FDA Inspection Under the QSIT Design Controls Subsystem," Washington, D.C. (www.AdvaMed.org).

FDA, August 1999, "Guide to Inspections of Quality Systems" (www.fda.gov).

Gopalaswamy, Venky and Justiniano, Jose M., 2003, *Practical Design Control Implementation for Medical Devices*, Boca Raton, FL: Interpharm/CRC Press.

Office of Health and Industry Programs, Division of Small Manufacturers Assistance, June 1996, *Investigational Device Exemptions Manual.*
CDRH, March 11, 1997, "Design Control Guidance for Medical Device Manufacturers.
ANSI, 1995, ANSI/ASQ D1160-1995, Formal Design Review.

chapter three

Six Sigma roadmap for product and process development

In Chapter 1 we mentioned that there has been tremendous focus on Six Sigma initiatives by many different companies in various industries over the past few years. This Six Sigma effort has resulted in improved product and process performance, improved supply chain performance, and so on, thereby clearly signaling that this approach can be used to achieve strategic business objectives. Most of the publications and books in the Six Sigma area, though good at explaining both technical and business aspects, are focused on applying this methodology for manufacturing or transactional processes. There are only a limited number of publications that focus on applying Six Sigma to design and develop products and associated manufacturing processes. To our knowledge, there are no books that specifically focus on applying Six Sigma to medical device design and development.

Chapter 2 of this book provided the readers with an overview of Design Control guidelines for medical devices. Elements of Design Control such as design plan, design input, and design output help the industry professional to understand what it takes to make the devices safe and effective. Quality system policies and procedures are implemented to ensure consistency in applying these regulations. However, these Design Control-related policies and procedures are usually not established on ensuring medical device manufacturers meet their non-compliance-related business goals. It can be argued that successful achievement of non-compliance-related business goals could be a derived benefit from successful implementation of Design Control policies and procedures. For example, it can be argued that successful implementation of design controls can result in a medical device that is cost-effective in addition to being safe and effective.

It is important that roadmaps are established to ensure that both compliance and non-compliance goals are met successfully. While there can be many non-compliance goals that a medical device manufacturer pursues, we focus on key product development-related non-compliance goals that we think are appropriate. So what are these key non-compliance goals that a medical manufacturer must focus on once a decision is made that a concept is going to be developed into a medical device?

- Designing, developing, and commercializing cost-effective devices that meet customer requirements consistently with extremely low variation
- Ensuring that the research and development-related resources are optimally utilized to commercialize these products as fast as possible
- Designing and developing effective and economical supply chain(s) that is (are) also safe and environment friendly

It is quite possible to visualize a medical device manufacturer having two distinct approaches to achieve these compliance and non-compliance goals, thus creating a "two-pile" approach. The manufacturer must pursue Design Control guidelines to meet compliance goals and may pursue a Six Sigma approach to meet non-compliance goals. It is not unusual to see that most device manufacturers treat Design Control requirements with extreme care and do everything possible to meet them. The same is usually true for non-compliance business requirements such as:

- Optimized project budget
- Schedule adherence to meet project completion dates
- Use of available information technology systems, and so on

However, when there are options, project teams usually take the path of least resistance in order to meet the above-mentioned requirements. Approaches such as Six Sigma methodology for product and process development may be treated as "optional," as shown in Table 3.1. It takes a lot more commitment from top leadership to emphasize the importance of Six Sigma as a roadmap as well as a management philosophy that can be made integral to the requirements mentioned in Table 3.1 below.

Table 3.1 Optional against regulatory requirements

	Compliance	Non-compliance
Requirements	Design Control	Company-specific requirements
Optional	N/A	Six Sigma

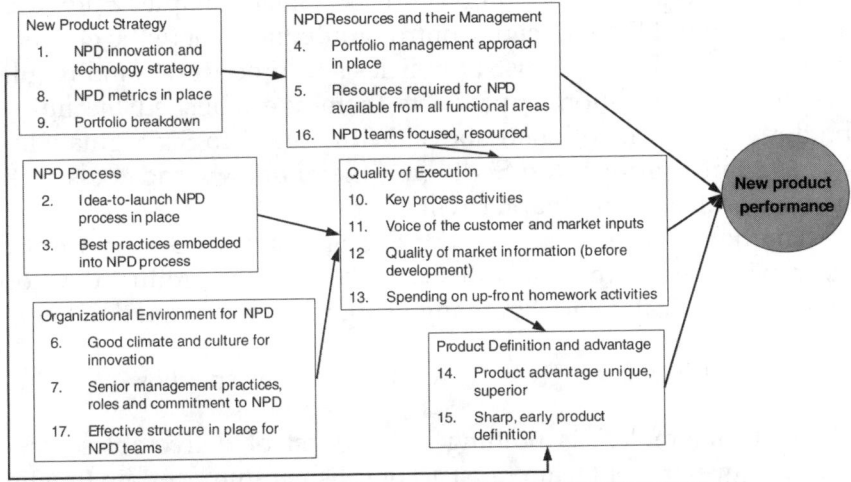

Reference: "Improving New Product Development Performance and Practices", APQC Best Practice Report, 2003

Figure 3.1 Best practices in NPD as presented by APQC.

Six Sigma approaches for design and development of medical devices will work best only when the framework for successful new product development (NPD) is understood.

Research done by the American Productivity and Quality Council (APQC) highlights the need for 17 best-in-class attributes for new product development. These attributes are further grouped into six different categories. The attributes, categories, and their linkages are shown in Figure 3.1.

From the figure it can be inferred that, in addition to meeting company goals, the following key characteristics must be present in any medical device company to ensure that the devices designed, developed, and released by the company meet or exceed customer wants and needs:

1. Presence of a business strategy leading to product portfolio
2. Presence of an effective organization climate and structure that includes but is not limited to cross-functional teams, management commitment, and innovation climate
3. Presence of an effective Design Control process

    4. Utilization of project plans with clearly identified milestones or deliverables

More specifically, of the six categories, it is safe to assume that both Design Control guidelines and Six Sigma approaches focus primarily on the following three categories: NPD process, quality of execution, and product definition and advantage. Chapter 2 provided an overview of FDA's Design Control guidelines where the elements of the "waterfall model," use of policies and procedures, and regulatory body classification of products thereby established the link to the three groups mentioned above. With regard to Six Sigma's link to these three groups, the Six Sigma methodology and tools to be discussed later in this chapter will establish it.

Another way to explain how Six Sigma and Design Control encompass these three categories is to characterize product development in a medical device company using the simple equation below:

$$\text{Deliverables} = (\text{what} + \text{why}) + \text{who} + \text{when} + \text{how}$$

where "Deliverables" is nothing but the list of deliverables that a product development team must accomplish within a certain timeline and investment, which will result in a successful medical device, that can either go to clinical trials or commercial market release. The term "(what + why)" stands for Design Control-related requirements that are usually found in company quality system policies and procedures. These requirements inform the medical device design and development teams on what needs to be done to get the product to clinical trials or commercial release to the customer. They also explain why these requirements must be met. The term "(what + why)" can also include the business needs, such as product target cost and scrap rate, which are expected from the product(s) that must be delivered by the product development team.

The term "who" in this equation is the project team that has the accountability to design and develop product(s). While the term "when" indicates the timeframe to deliver products to clinical trials or commercial release, the term "how" points to the various engineering and statistical tools and methodologies that are needed to successfully design and develop medical devices.

For example, if one of the deliverables from the design team is a risk analysis summary report, then the above equation might look like:

Risk analysis summary report = (risk analysis standards
[ISO 14971] + regulatory agency filing requirement) + project
team + before regulatory submission + FMEA

While Chapter 2 focused on the "(what + why)" term in the above equation, this chapter will mostly focus on the "how" term. Specifically, this chapter will focus on providing an overview of the tools and methodologies that can be brought under the umbrella concept called Design for Six Sigma (DFSS). Details on other terms are beyond the scope of this book and can be obtained through other relevant publications.

The implementation of FDA's Design Control guidelines by medical device manufacturers almost always led them to implementing a design and development process. This process usually includes four or five stage/toll/stage gates and incorporates FDA's Design Control guidelines. As a new product is designed and developed, according to Prof. Nam Suh of MIT, the new product development process takes the design team through four different domains: Customer, Function, Design, and Process.

In medical device design and development, it is safe to assume there is a fifth domain that is present before product development enters the customer domain. We call this the "innovation domain." This is because medical device companies constantly must innovate in order to survive over the long run. Unlike many other industries, a large portion of product ideas in the medical device industry comes from external sources such as device users and universities. These ideas as well as those that are generated internally must be evaluated and acted upon to improve the companies' intellectual property. Patents and trade secrets are a few of the measures used to keep track of the strength of the intellectual property.

The innovation domain can also be viewed as something that is present in the other four domains due to the possibility of innovation that can occur within these domains. Since the scope of this book is limited to Design Controls and Six Sigma, we will not focus on the up-front innovation domain as it is usually outside the scope of FDA's Design Control guidelines. We will, however, focus on the innovation that is embedded in the other four domains.

In this chapter we will introduce the concept of Six Sigma for product and process development, explain different approaches needed to effectively apply Six Sigma to product development, and provide an overview of various process and quality improvement tools that are part of the Six Sigma approach.

Six Sigma has been in existence ever since it was used in Motorola in the early 1980s. However, General Electric's past chairman and CEO Jack Welch is widely credited for fueling the move by many industries to apply Six Sigma principles over the past decade. While the initial emphasis of Six Sigma was in applying it to manufacturing, recent conferences in Six Sigma tend to focus more on applying Six Sigma to product development and to other functional areas and processes outside of manufacturing. Companies such as GE, Allied Signal, and Raytheon have successfully implemented Six Sigma methodology for designing and developing products. The Six Sigma methodology used to design and develop products is commonly referred to as Design for Six Sigma. There are many acronyms that are used to describe the different stages or phases within DFSS. Two of the most popular ones are:

1. DMADV → Define, Measure, Analyze, Design, Verify/Validate
2. IDOV → Identify, Design, Optimize, Verify/Validate

Fundamentally these two are the same. They both focus on the following key activities within new product development:

- Defining or identifying customer wants and needs
- Measuring and analyzing these customer wants and needs to develop key functional requirements
- Designing a product (which includes its packaging) and its associated manufacturing processes to these design requirements
- Verifying and validating both the product and its associated manufacturing processes

It is a well-accepted notion that the concept of Six Sigma, when implemented properly in the design and development process, will improve a company's top line due to increased sales and reduced product development cycle time. However, we have also observed that there is some hesitation among product design and development personnel in adopting Six Sigma for design and development. The situation can be slightly worse in medical device companies, since the recent introduction of FDA's Design Control guidelines has already created the impression among product development personnel that these guidelines might limit their ability to innovate. Asking them to adapt Six Sigma approaches can almost create resentment. Is this a fault of the DFSS approach? We most certainly think it is not.

The fault usually lies in the deployment of these approaches. We strongly believe that methodologies such as Six Sigma must be integrated with stage-gate processes for product development to result in an enhanced stage-gate process. This will eliminate the "two-pile" approach mentioned earlier. We believe that it can be accomplished by using the simple equation that we presented earlier in this chapter for every deliverable.

It is a well-known fact that many of the Six Sigma tools are not new. The discipline of quality engineering has always emphasized the use of these tools in product design and development. However, what is new is the application of "system thinking" to use these tools. What do we mean by "system thinking"? It is the integrated application of these tools to flow-down requirements and flow-up capabilities to design and develop products.

Figure 3.2 is a visual representation of this approach, and it clearly highlights the benefits of simultaneous consideration of requirements and capabilities throughout the new product design and development process. While requirements for the product design and development "flow down" from new product development to the supply chain process, the capabilities of the supply chain process "flow up" to the new product development process, thus creating an environment and an effective approach where both compliance and non-compliance goals can be met. For example, if the supply chain process of a medical device company has competency in manufacturing mechanical parts and the new product development group(s) is (are) focused on new product designs that include electronics and software technology, then supply chain group(s) should be involved in both strategic and tactical capability discussions early in the product development process.

DFSS tools can be mapped to the four domains indicated in Figure 3.3 to develop a handy illustration such as the one in Figure 3.4. This list of tools in the figure does not necessarily mean that all the tools are applicable for all device design and development projects. It also does not mean that these are the only tools that are applicable for medical device design and development projects. We will provide an overview of some of the key tools along with key activities that should be included as well as excluded during the application of these tools to make them more effective. We provide them in a simple table format of "do's" and "don'ts." Just for clarity, we want to point out that the readers should include the words "Do" or "Don't" before each item in the tables so that they make sense. For

Flow-down requirements

| Customer wants and needs | Medical device functional requirements | Design and develop medical device, its packaging and manufacturing processes | Verify and validate product and manufacturing processes |

Flow-up capabilities

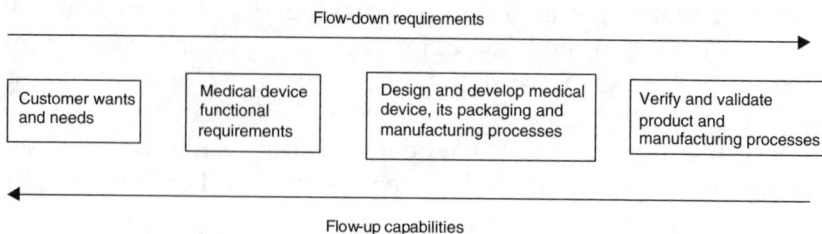

Figure 3.2 Medical device design and development domains.

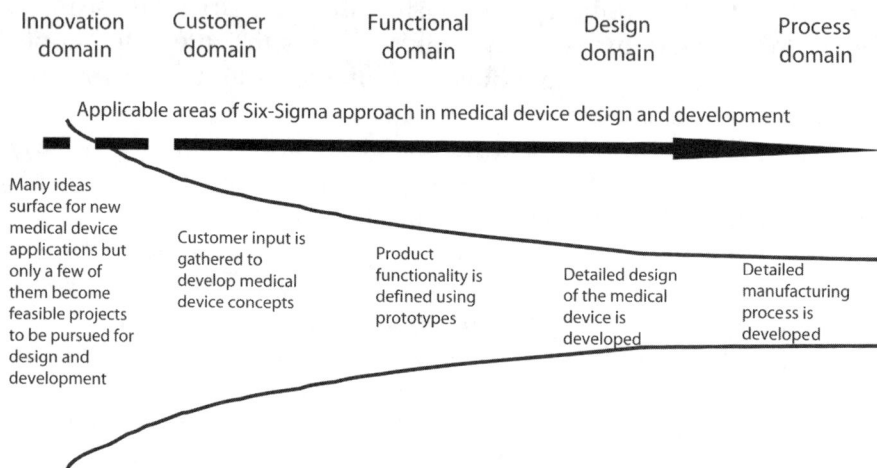

| Innovation domain | Customer domain | Functional domain | Design domain | Process domain |

Applicable areas of Six-Sigma approach in medical device design and development

Many ideas surface for new medical device applications but only a few of them become feasible projects to be pursued for design and development

Customer input is gathered to develop medical device concepts

Product functionality is defined using prototypes

Detailed design of the medical device is developed

Detailed manufacturing process is developed

Figure 3.3 Design for Six Sigma approach.

more details on how to apply these tools, we refer the readers to the references cited at the end of this book.

It is important that the product development teams create a plan up-front on what tools can be applied based on the deliverables. If the teams have difficulty in coming up with such a plan, we recommend that the teams consult with a Six Sigma expert. It is acceptable to have an outside expert or consultant help the team, but in order to sustain the effectiveness and efficiency improvement that will result in deploying DFSS, medical device companies must develop internal experts and make them available to other teams.

These experts can help in identifying the training and coaching necessary as well as in planning on how to make them available to the team just in time. We strongly believe that unlike training for a Six Sigma product and process improvement methodology such as DMAIC, training for the DFSS methodology is not effective when offered in "waves," but it will be effective if offered when the team

| Innovation domain | Customer domain | Functional domain | Design domain | Process domain |

Applicable Six Sigma tools in medical device design and development

| Concept Generation (TRIZ etc.) DOE | VOC QFD Financial models | DfX Pugh Matrix Reliability FEA DOE System FMEA | DOE Simulation Statistical tolerancing Design FMEA | SFC Validation DOE/ Response Surface Process FMEA |

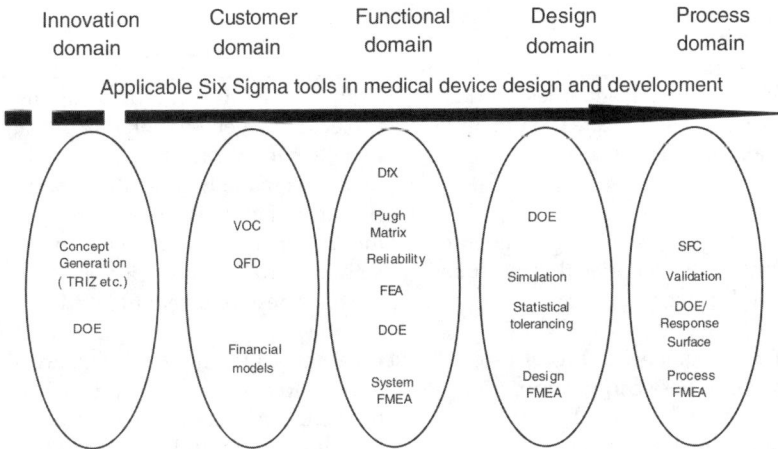

Ref: Larry R. Smith, Ford Motor Company, "Six-Sigma and the Evolution of Quality in Product Development", Six Sigma for Chemicals & Pharmaceuticals, IXPERION Annual Summit, 2002

Figure 3.4 Six Sigma tools mapped to product development domains.

needs it. In other words, project teams cannot be trained one week at a time or project team leaders for all active projects cannot be trained at the same time. The reasons for this include a longer timeframe required to complete design and development of a medical device compared to a process improvement project and the variety of expertise needed (FEA, Statistics, Lean Manufacturing, Process Technology) by each project team to successfully develop products. We also believe that design reviews are effective mechanisms available to ensure that the teams utilize the DFSS approach.

Customer wants and needs or customer domain

Project planning

A medical device design and development project is typically initiated after some successful analysis (outside of Design Control requirements) of innovative concept(s) and when there are indications that the new device can be successfully commercialized. In some cases, success in clinical trials can be a key milestone prior to commercialization due to the nature of the product. Best-in-class medical device companies typically have fully formed cross-functional project team(s) at this point. Project planning activity in DFSS is the tool that captures the "who" and "when" parts of the deliverables equation mentioned earlier in this chapter.

Table 3.2 Tips to improve project planning

Do's	Don'ts
Identify all possible activities needed and periodically update the timelines reflected in the software.	Blindly follow timelines reflected in the project management software, as it usually cannot capture many subtle decisions made that result in parallel activities that occur in reality.
Refer to company operating procedures to identify deliverables that must be included in the plan.	Underestimate the time required for activities that must be performed at contract design or manufacturing facilities.
Plan for contingencies during project execution. These can include environmental, political, and organizational contingencies.	Forget to include time and resources needed to assess (and implement if necessary) an acceptable quality system both in-house and outside facilities.
Identify the scope of the project up-front so that only relevant activities are included in the plan.	Forget to include packaging, transportation, storage, and sterilization activities. These areas are often treated as an after-thought, which usually results in a lot of wasted effort after product launch. This is due to problems in the field.

These teams are funded appropriately to achieve their milestones. The teams usually start their activities with a project plan, which is represented at a high level using a Program Evaluation and Review Technique (PERT) or GANTT chart. Textbooks in project management can provide guidance on these terms and how they are used in project management. Additional tips to improve project planning are shown in Table 3.2. It is not uncommon to observe project teams using software such as MS Project® to capture all the planned activities and resources needed to make the project successful.

Here are some of the typical questions that a plan shall answer:

- Do we have a complete team? If not, when can we expect additional or specialized resources to be available? (e.g., When does the team need a specialist in software reliability?)
- Do we have independent peer reviews of the project scheduled at every stage-gate?
- Does everybody on the team or supporting the team know what is expected from them? Do they know their due dates and deliverables?
- When are the team's decisions due and what decisions are expected from the team?

- Are we going to meet the business targets (e.g., schedule, cost, risk, and reliability levels)?
- Where are we on this project?
- Where, when, and how can top management help?
- What is the critical path? How can we ensure that falling behind schedule is not compensated with compromised quality and reliability?
- Do the list of planned activities include addressing all quality systems requirements such as:
  - Bio-compatibility
  - Product risk analysis
  - System FMEA, Design FMEA, Process FMEA
  - Environmental impacts
  - Component qualifications
  - Accelerated aging
  - Pre- and post-sterilization correlation analysis
  - Process stability
  - Product stability (shelf life)
  - Packaging stability
  - Validation of test methods, GR&Rs, measurement capability analysis, etc.
  - Process and design validation

Stakeholder requirements identification and analysis

This analysis captures the Critical-to-Business (CTB) requirements and helps the teams identify key stakeholders that need to be "managed" during the project. It is a simple DFSS tool that must be completed just before or after the development of the project plan. Table 3.3 is a sample template for this tool.

It must be noted that not all stakeholders' wants and needs are directly linked to the product that is developed by the project team. Stakeholders in a medical device company can include senior management in various functional organizations such as quality, regulatory, finance, operations, and information technology. Identifying these key groups and planning activities to mitigate any potential surprises will go a long way in the team's ability to complete the project successfully.

Teams typically fail or get caught in organizational politics if CTB needs are not accounted for in their project planning efforts or, for that matter, in their product design. How often have you heard the following questions or comments?

Table 3.3 Template to complete stakeholder analysis

Stakeholder	Stakeholder requirements	Current support level (1 = low; 5 = high)	Desired support level (1 = low; 5 = high)	Action plan	Team member responsible

- If only the project team consulted the software validation engineer up-front, the project would not have been delayed.
- Why didn't they include a reasonable product target cost in their design efforts? Now we have a product that costs a lot to manufacture.
- Why didn't the new product development team consider long-term manufacturing needs when they selected the process technology to make the product?
- Why didn't the new product development team select a supplier that has GMP-compliant facilities?
- Regulatory professionals were recommending that the team consult with appropriate regulatory agencies as early as possible on certain decisions but to no avail.
- I can't give up my quality assurance resources just because your project team failed to plan up-front.

In addition to accounting for CTB needs, project teams must identify appropriate team members to manage key stakeholders (see Table 3.4). Critical-to-Business wants and needs can include deliverables such as:

1. Acceptable scrap rate at launch
2. Target product cost at launch
3. Project costs not exceeding allocated budget
4. Project documents in standard information technology format
5. Human resources-related deliverables
6. Utilization of approved component suppliers and original equipment manufacturers

Table 3.4 Tips to improve stakeholder identification and analysis

Do's	Don'ts
Identify all key stakeholders who will be impacted by the project and all key activities that must be included in the project plan to "manage" these key stakeholders.	Plan to spend equal amounts of time to manage all stakeholders. This can result in reduced focus on project deliverables.
Prioritize stakeholders based on their impact to project timeline, budget, quality, etc. This will help manage the project timeline and budget better.	Forget to include requirements from other project teams that might be planning to use output from your project.
Include FDA and other regulatory agencies as stakeholders. Please note that as a conservative approach, these agencies can be considered as customers.	Forget to resolve any conflicts between stakeholder needs and customer needs. (For example, a much higher profit margin expectation that can be impacted by a growing customer expectation of an inexpensive but high-quality product).

SIPOC

SIPOC is an acronym for "Supplier, Input, Process, Output, Customers." It is a simple visual representation of the inputs needed by the design and development team from its suppliers (e.g., stakeholders, component or finished goods suppliers) to deliver specific outputs to its customers (e.g., patients, surgeons, stakeholders).

The main use of this tool in the customer domain is to identify customers and suppliers that are needed to successfully complete the product development project (see Table 3.5a). The "Process" portion of the SIPOC at this stage is usually a high-level map of the stage-gate process that is followed by the team. Note that there are other process mapping tools, such as "Value Stream Mapping (VSM)" and "Functional Mapping," that are useful in medical device design and development. These tools are useful during the later stages of the stage-gate process. Ideally we prefer the project teams to use SIPOC and other process mapping tools.

Project risk analysis and management

Project risk analysis and management is an important tool that is often overlooked by most product development teams. The fact that a medical device development team is usually involved in delivering a safety-critical product to the market as early as possible poses a

Table 3.5a Tips to improve SIPOC tool application

Do's	Don'ts
Focus on identifying all *key* suppliers, inputs, outputs, and customers. Refer to company's operating procedures for stage-gate process or FDA's Design Control deliverables.	Spend time to analyze the needs of all suppliers and customers at this point. Spend time trying to figure out the details behind the "Process" in SIPOC. Unless there is an already existing process, this will be a wasted effort at this point.

Figure 3.5b Sample project risk analysis decision matrix

Severity ⇓	Likelihood		
	Low	Medium	High
High	Potential show-stopper	Show-stopper	Show-stopper
Medium	Potential show-stopper	Potential show-stopper	Show-stopper
Low	Acceptable level	Acceptable level	Potential show-stopper

compelling argument to use project risk analysis and management as a tool.

Project risks are identified and they are analyzed for their effect on project cost, time, quality, and so on. These risks usually are business, organizational, competitive, financial, legal, and regulatory in nature. They are then ranked based on the severity of the effect and the likelihood of their occurrence. The scale used for severity and likelihood can be qualitative such as "low, medium, high" or quantitative such as "1 to 5." The intent is to identify and eliminate any "show-stoppers" as well as minimize other potential risks. Project risk analysis is a simple but effective tool. Figure 3.5b illustrates a project risk decision matrix, which can be used to make decisions on the level of each project risk. Some tips identified in Table 3.6 should help in analyzing project risk.

Voice of the customer (VOC)

Once a project plan is developed and CTBs are identified, the project team's next step is to find out what the customer wants and needs. The customers for a medical device include surgeons, clinicians, physician assistants, and patients. They can also include biomedical engineers, hospital materials managers, and technical personnel involved

Table 3.6 Tips to improve project risk analysis and management

Do's	Don'ts
Focus on identifying all key project risks including high-level technical risks.	Get bogged down in details or in finding immediate solutions.
Create a scale to consistently rate the severity and likelihood of potential project risks.	Try to assign resources to solve all categories of risk.
Focus on business and organizational "show-stoppers" first. Use stakeholder analysis in conjunction with this analysis.	Start to plan mitigation steps prior to ranking and understanding the risks.

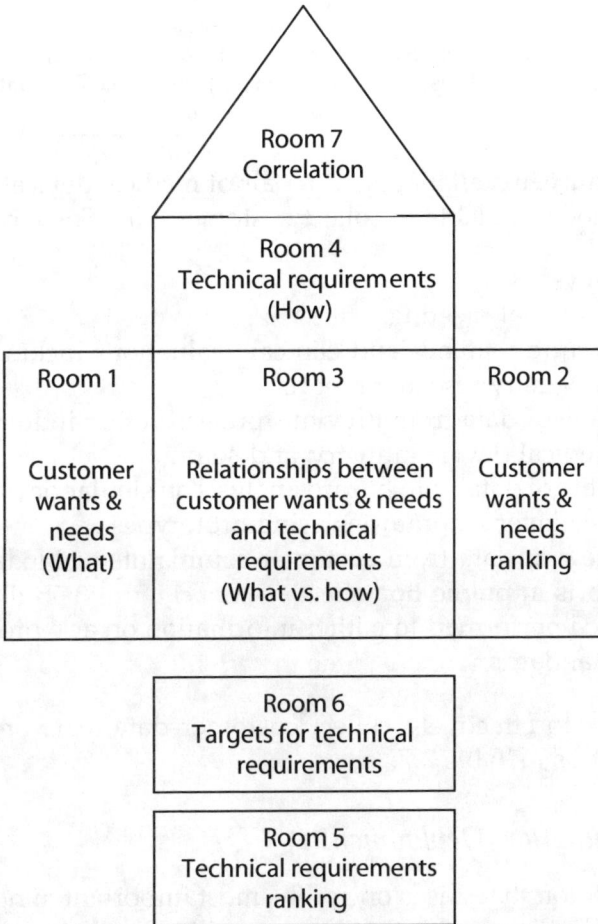

Figure 3.5 House of Quality.

Table 3.7 Tips to improve customer data collection

Do's	Don'ts
Plan up-front on which of these tools are necessary (e.g., complaint data) and which other ones will be used to collect data on customer wants and needs.	Depend on reactive tools (e.g., complaint data review) more than proactive tools (e.g., survey).
Segment customers based on their importance to the performance of the product.	Treat customers and stakeholders in the same group.
Perform in-depth analysis of customer complaints and MDR data to identify opportunities for improvement.	Rely on only one type of data collection technique (e.g., survey). Depending on the number of responses or the nature of response required, more than one technique is usually applicable.
Utilize quantitative market research including conjoint analysis.	Assume only marketing and sales experts need to collect data from customers.

in maintenance, installation, and repair of medical devices. There are many methods available to collect customer data. Some of them are:

- Surveys
- Focus panel meetings
- Literature searches and clinical evaluations including review of university research reports
- Review of data from previous projects, other industry outside of medical device industry, and so on
- Review of data from hazard analysis on similar devices, animal and cadaver laboratories using prototypes
- Review of data from customer complaints or Medical Device Reports in public domain
- Search performed to gather information on any product-related standards

In order to effectively collect customer data, it is important to follow the tips in Table 3.7.

Quality Function Deployment (QFD)

It is our opinion that this is one of the most important, if not the most important, Six Sigma tools that must be utilized by the medical device community during product design and development. QFD is a simple but tedious tool that provides the project team the needed link between customer wants and needs and design input. As such, use of this tool naturally follows the data collection methods mentioned earlier. QFD can also be considered as a process with inputs and outputs.

The term "QFD" is also used interchangeably with the term "House of Quality" (HOQ), due to its shape. There are typically seven rooms in the House of Quality as shown in Figure 3.5. Each room captures a specific element in the QFD process. They are customer wants and needs (room 1), customer ranking (room 2), technical or Critical-to-Quality (CTQ) requirements (room 3), relationship matrix (room 4), technical requirements ranking (room 5), targets for technical requirements (room 6), and correlation of technical requirements (room 7). When all the rooms are filled and understood, a product development team should be fairly equipped with the knowledge of how the design effort should progress. It must be mentioned that we have seen HOQ without room 5 due to lack of competitive benchmarking data. This led us to coin the term "Townhouse of Quality" for this modified HOQ. Is this a problem? We think so, since in new medical device development it is always a good practice to understand the technical requirements for a competitive or predicate device before setting targets and specifications for CTQ requirements.

Engineers and other professionals on the product design and development team can use this tool to prioritize the Critical-to-Quality requirements for the medical device. The Critical-to-Business requirements are usually not considered in QFD development. However, there is no rule that is stopping anyone from creating a separate House of Quality for CTB needs or adding these requirements to the CTQ list in the House of Quality. Additional tips are listed in Table 3.8.

Table 3.8 Tips to improve Quality Functional Deployment

Do's	Don'ts
Group similar wants and needs such that they can be addressed better.	Address every requirement that is listed in QFD unless necessary.
Collect customer input to complete room 2.	Complete every room without clearly understanding what is being input.
Try to capture as many customer wants and needs as possible in QFD.	Assume all CTQ needs are important. If they don't strongly correlate with customer wants and needs, eliminate them.
Prioritize CTQ before finalizing the list of requirements.	Develop a QFD in isolation. This activity must involve the entire product development team and supporting organizations.

Table 3.9 Tips to improve functional block diagrams

Do's	Don'ts
Create separate block diagrams for each intended function of the medical device	Forget to include human interfaces that are needed to make the device function (e.g., pushing a button to activate a dental cleaning device).
Show any planned redundancies in the system.	Create functional block diagrams for potential unintended use of the device.
Try to assign reliability or failure rate values for each block based on existing or similar devices.	Check for the presence of any block in other block diagrams before eliminating.

Please note that we did not provide details on how to fill each room in the House of Quality since it is outside the scope of this book. We refer you to many books published on Six Sigma that usually have a good description with examples on the application of QFD (see Appendix 3).

Product functional requirements or functional domain

Functional block diagrams

Functional block diagrams are useful to illustrate how the different functional concepts (or high-level designs) of the medical device will work together to make the medical device perform its intended function. We strongly encourage the readers to refer to Table 3.9 and our book on design control for details.

Theory of inventive problem solving (TRIZ) and other concept generation tools

Pronounced "Treez," TRIZ is an approach to generating innovative concepts that was developed in Russia by Dr. Altschuller. Though we have not seen or heard of widespread use of this tool in the medical device industry, we believe it is a useful tool when the medical device that is being developed is complex or when there are many conflicting CTQ requirements. For example, if one CTQ requires the medical device to be flexible and another CTQ requires it to be rigid, TRIZ can help the designer to identify a solution domain where this conflict can be addressed better. Many other concept generation tools, such as "brainstorming" and "solution mapping," can be utilized to develop concepts for design as well as supply chain elements. Proper use of these tools can result in not only better concepts but also reduced cycle time. For additional tips, refer to Table 3.10.

Table 3.10 Tips to improve concept generation tools

Do's	Don'ts
Use one or two concept generation tools depending on the complexity of the problem or conflicts that need to be resolved.	Forget to include concepts that were considered in the past but shelved due to other reasons.
Focus on generating at least three or four high-level design concepts first before evaluating them.	Include management in concept generation sessions. This can result in a non-optimum solution.
Include customers (surgeons, hospital support staff) whenever possible.	Forget to have a clear understanding of intellectual property constraints.

Reliability allocation

Once functional block diagrams are created, it is recommended that the design team allocate reliability goals to each and every block based on expected overall system or medical device reliability. A reliability specification should include the medical device's intended mission and a reliable metric that will be used to measure the reliability. The overall medical device reliability goal can also be set based on expected non-safety-related field failure rates or allowable defect rates. For example, a reliability goal for an endoscope can be as follows: The endoscope system must operate without failure for 10,000 uses under normal operating conditions and must provide clear view of internal organs, with 99.999% clarity. For example, if a predicate endoscope had a field failure rate of 1 in 1000 units when used as intended, then the reliability of a new device can be approximated to be equal to or better than $1-e^{-(t/1000)}$, where t is the intended life of the new endoscope (10,000). This will result in a reliability goal of 0.99996. Reliability allocation then takes into account how the various blocks are connected — in series or parallel or combination. In addition to Table 3.11, for details on reliability allocation and risk analysis, we refer the reader to our book on design control implementation.

Design considerations for human factors, manufacturing, supply chain, and environment

The ever-increasing scrutiny of medical devices by the regulatory agencies, the public, and the media as well as the increasing awareness of customers through competitive advertising and the Internet are forcing medical device manufacturers to pay more attention to human factors and environment in the device design and development. In the United States, the regulations and guidelines published

Table 3.11 Tips to improve reliability allocation

Do's	Don'ts
Start with the system reliability in which the medical device will be a major "subsystem." Other subsystems include device packaging, interfaces during its operation, etc.	Forget to include reliability effects due to chemicals (e.g., pharmaceuticals) or software present either as part of the medical device or during use of the medical device.
Analyze any existing complaint and Medical Device Report (MDR) data to identify possible reliability goals.	Allocate reliability evenly to all planned components in the medical device. Determine their allocated reliability based on risk.

Table 3.12 Tips to improve design considerations for human factors, manufacturing, supply chain, and environment

Do's	Don'ts
Group similar wants and needs such that they can be addressed better.	Overlook or compromise these design considerations.
Collect customer input on these design considerations and incorporate them into QFD.	Fail to create database(s) that can be used by other teams to standardize these requirements.
Depending on the need, involve other experts when developing these requirements.	Design multiple, complex, or long supply chains to save time during design and development.

by the FDA and Environmental Protection Agency (EPA) provide the bulk of the requirements that a project team has to include in their designs. For other countries, we recommend the project teams consider regulations and guidelines from countries where the product will eventually be released.

The presence of competition and increasing expectations from regulators and investors have forced medical device manufacturers to address supply chain and manufacturing development early in the device design and development. This trend helps the implementation of DFSS since it helps to incorporate these requirements either in Critical-to-Quality or Critical-to-Business needs for the project team to be successful. If suggestions similar to those in Table 3.12 are not followed, we are almost certain that companies will create more "opportunities" for improvement or "DMAIC" projects in the future and spend resources on fixing or improving product-related problems. We believe that this dimension of Six Sigma in design and development is not emphasized enough by other publications on this topic. We also strongly believe that all four of these design considerations will eventually decide whether a new medical device can sustain its position in the marketplace over time.

Product and process design or design domain

Pugh or prioritization matrices

Pugh or prioritization matrices are simple Six Sigma tools that help the medical device design and development teams select a final design concept that can be built into a prototype design for further testing and development. While the Pugh matrix evaluates new high-level medical device designs relative to a baseline (predicate) or benchmark design, the prioritization matrix evaluates new designs to an absolute scale of weighted requirements. The criteria used to evaluate high-level designs are nothing but the CTQ and CTB needs described earlier. While there are many ways to effectively use these tools, the tips given in Table 3.13 should help in the design of medical devices.

Measurement systems analysis (Gage R&R)

This one of the key tools within DFSS, since lots of decisions about key product characteristics (CTQ) are made during medical device design and development using measurement systems. Measurement systems for attribute and variable characteristics need different analyses to be performed to assess their repeatability and reproducibility. Most of the off-the-shelf (OTS) statistical software such as Minitab™, Statgraphics™, and Statistica™ have capabilities to perform this analysis. It is highly recommended that, as a minimum, CTQ requirements-related measurement systems are validated. This is especially true when products from newly acquired companies are integrated. For additional guidance, please refer to Table 3.14 or any textbook on measurement systems analysis.

Table 3.13 Tips to improve Pugh or prioritization matrices

Do's	Don'ts
Include both CTQ and CTB needs to select the high-level design concept that needs to be further developed.	Limit the high-level design concepts to what is available. The CTQ and CTB needs can be addressed with a combination of these designs.
In case of conflicts, CTQ should be given more importance than CTB.	Finalize concepts without the project team's involvement.
Where possible, select the top two high-level design concepts to minimize risks.	Forget to verify CTB and CTQ needs addressed by other project teams for concept selection.

Table 3.14 Tips to improve measurement system analysis

Do's	Don'ts
Where possible, use validated existing or industry-standard measurement systems. This will help with FDA or other regulatory agency submissions or during audits.	Blindly assume that product development experts know how to measure new designs. A weak test method can make a good product to be rejected. (Think of the consequences it can have on FDA submission timelines!)
Select variable characteristics over attributes for better inference from results.	Use gages whose repeatability and reproducibility value is greater than 30 to 40%.

Product risk analysis, including Failure Mode and Effects Analysis (FMEA)

FMEA is one of the tools that is readily accepted by product development teams because of its simplicity as well as the necessity to perform medical device risk analysis prior to regulatory submissions. As indicated in Table 3.15, FMEA sessions are effective when customers are included, someone knowledgeable of the tool facilitates it, and the session focuses on identifying key failure modes (as against all failure modes).

We have provided a detailed review of this tool in our book on design control. We also have referenced appropriate international standards that are applicable to performing risk analysis for medical devices.

Table 3.15 Tips to improve product risk analysis

Do's	Don'ts
Include the entire design team, customers, and suppliers in design FMEA sessions. Ensure that someone who knows the tool facilitates the FMEA session.	Try to solve failure modes as they are identified during an FMEA session. This is the number one reason why many design and development engineers feel that FMEA sessions are not productive.
Consider the effects of failures on CTQ needs identified earlier in QFD followed by CTB needs. This will help mitigate risks that have potential impact on CTQ first.	Try to identify all potential failure effects and causes for all potential failure modes. Start with five to ten key potential failure modes with no more than three effects for each mode.
Use area charts based on severity and likelihood to evaluate the risk landscape in addition to tables.	Forget to update FMEA whenever there is a significant change to the product during design or before a design review.

Table 3.16 Tips to improve modeling and simulation

Do's	Don'ts
Start this activity as early as possible in the design process.	Create simulation models for simple systems with few inputs and outputs.
Ensure that data can be readily pulled into the models. Lack of this ability is the number one reason for failure to implement modeling and simulation in the design process.	Try to manage simulation models as a team. It works best if the team provides input to a specialist who is capable of running these models.

Modeling and simulation, including Finite Element Analysis (FEA) and Discrete Event Simulation

One of the major reasons for the advances in product design and development in the automotive and aerospace industries is the ability to learn about products and associated manufacturing processes before a single prototype is developed. Modeling and simulation allows new product and process development teams to virtually design and develop products and processes to understand their weaknesses and bottlenecks. This in turn can help in making improvements to optimize to the product and process designs. There are many different types of modeling software available to accomplish these. They include but are not limited to Finite Element Analysis, Plastic Mold-Flow Analysis, Discrete Event Modeling and Simulation, and Transfer Function Analysis. Effective use of modeling, as shown in Table 3.16, can help the design team avoid future problems.

In-depth use of descriptive statistics and experimentation

This includes confidence intervals, hypothesis tests, continuous and attribute response inferences, continuous and attribute response comparison tests, correlation and simple linear regression, analysis of variance, and multiple regression. While designing a medical device, project team members collect data through various means such as animal laboratory tests, bench-top tests, and computer simulations. Statistical analysis is key in helping the team members make the correct decision with a high level of confidence. In addition, we suggest that due consideration is given to the items listed in Table 3.17.

Design of Experiments (DOE) and optimization

Design of Experiments is probably one of the most important Six Sigma tools in the development of medical devices, and it is also one

Table 3.17 Tips to improve statistical analysis during product development

Do's	Don'ts
Consult a subject matter expert before making decisions.	Blindly accept output from statistical software. Always have an expert interpret the results and get their signature of approval to avoid future problems.
Have "criteria for success" so that decision making is simplified.	Use statistical software without completing some basic "software validation" activities for your company.
Use one confidence level (usually it is 95%) for all analyses for consistency.	Use electronic spreadsheets unless they are verified.

of the least used in the design of medical devices. This aspect always baffled us since this tool can help the designers and design engineers understand key aspects of the medical device and its packaging with fewer experiments than what is traditionally done. This also means that less time and money can be spent during design and development, which is something management always wants. When DOE is utilized, the preferred method of practitioners is classical DOE and not Taguchi approaches. We have found various reasons for this including familiarity with the tool and lack of appreciation or understanding of when the Taguchi approach is applicable. Irrespective of the DOE approach used, follow the interventions in Table 3.18.

Statistical tolerancing

Statistical tolerancing of subsystems and subassemblies and components based on overall product design dimensions must be done up-front in medical device design and development to ensure proper medical device form, fit, and function. Since product performance depends on robust design and robust manufacturing processes, all the learning must occur upstream in the design cycle. Process design and development is one area where less attention is paid during design and development. Since product design engineers are focused

Table 3.18 Tips to improve Design of Experiments

Do's	Don'ts
Use DOE as early in the design process as possible.	Forget to qualify measurement systems to be used prior to running the experiments.
Perform a confirmation run to verify if the optimum input conditions derived result in expected response(s).	Forget to include interaction effects in addition to main effects while analyzing the response(s).

Table 3.19 Tips to improve statistical tolerancing

Do's	Don'ts
Use historical data from production if existing components are used in new designs.	Forget to perform a worst-case analysis for both linear and non-linear tolerance stack-ups.
Use Monte Carlo simulation using software for complex geometry.	Forget to select a logical starting point (e.g., one side of an unknown gap dimension) for tolerance stack-up analysis.

on the deterministic design features of the medical device and since there is severe time pressure to release the products due to competition, they often pay little attention to understanding the probabilistic variation (raw materials, production) that occurs during day-to-day manufacturing. DFSS uses tools such as Monte Carlo simulation to understand this variation. In addition, considering that the life cycle of a medical device in the marketplace is much shorter compared to pharmaceuticals and that they are not high-volume products (> 1,000,000 units each year), it is not easy to understand potential variations in the subsystems or components.

It is a well-known fact that variation in production is almost inevitable. Lack of sufficient volume coupled with poor part tolerancing will only magnify this variation, since it will almost always lead to lot of "fire-fighting" (production problems leading to more scrap, customer complaints, line shut-downs), thus wasting lots of precious resources.

To mitigate risks posed, we suggest, as listed in Table 3.19, that product development teams either consider historical production data if existing parts are used in the design or use Monte Carlo simulation to generate production data to understand potential variation. This data can be used to perform "worst-case" or "root sum of squares" tolerance analysis to detect non-linear and linear variation build-ups in subsystems and components. Software such as Crystal Ball™ and @Risk™ can assist in performing statistical tolerancing.

Reliability testing and assessment: overview

A medical device that is designed and developed must be tested *in vitro* or *in vivo* prior to releasing the product for commercial use. As often is the case with medical devices, there are little to no redundancies in the product to increase reliability unless the device is more of a "dynamic" capital equipment such as computerized tomography or blood glucose monitors, compared to "static" devices such as

Table 3.20 Tips to improve reliability testing and assessment

Do's	Don'ts
When performing a reliability test, create a test protocol and ensure sufficient test samples, test methods, animal models, and trained test personnel are available before testing begins.	Stop the test after only two or three failures. This is especially true if the failure modes are different.
Stimulate failures by increasing the stresses on the medical device even if they are beyond what the product would normally experience in actual use.	Assume that the reliability (life) test data is normally distributed. Use Weibull distribution initially to fit the data and try other distributions if not successful.
Track reliability growth of the medical device design if there are many design iterations.	Test a medical device without any applied stress. These "success tests" (because the products will mostly pass the test) will often end in product failures in actual use.

hospital beds. A design engineer is always challenged with designing devices with fewer but more reliable and cost-effective components.

Once these designs are completed and frozen, it is necessary to verify that the design performs as intended. While product performance can be simulated and evaluated using computer software, our focus in reliability testing is on performing tests in a laboratory or clinical situation. Protocols are written and executed to generate reliability data. To do that, products are tested until failure occurs or until a predetermined number of failure units are observed. This data must be analyzed to know how reliable the device is. We have provided details on reliability testing and analysis in our design control book. In addition to Table 3.20, other textbooks in reliability can also help the readers in understanding and applying these techniques.

Verification and validation or process domain

Response surface methodology (RSM)

In the process domain of the DFSS approach, medical device manufacturing processes are fully developed, qualified, and scaled-up for commercial production. Note the use of the word "fully" in the previous sentence. This is due to the reality that most of the manufacturing process designs are performed in parallel to the medical device design activity during the design domain. We discussed this in the Statistical Tolerancing section earlier.

Table 3.21 Tips to improve RSM

Do's	Don'ts
Screen process variables first to narrow them down to a meaningful number and then optimize them using RSM.	Try to optimize every response variable. Always use a risk management approach to identify and prioritize critical response variables.
Include RSM as part of Operational Qualification (OQ) phase of validation. This will help in establishing the process window for regular production.	Blindly follow output from statistical software even if the software is validated. Try to understand if the response surface model makes engineering or scientific sense.

Once manufacturing and assembly processes are fully developed they must be qualified. Process validation-related QSR requirements must be met before commercial products are released. In the process domain, Design of Experiments are performed to challenge the process and to establish proper "process windows" to enable day-to-day production. If the medical device is developed to begin a clinical trial, the manufacturing processes must be verified.

Response Surface Methodology is a DFSS tool where optimum input or process conditions are established for the response required. For example, if the response required is peel strength for packaging, the input factors that need to be optimized can include pressure and temperature. Table 3.21 contains tips to improve RSM.

Control charts

Once the manufacturing and assembly processes are optimized, it is necessary to establish control or precontrol limits so that the quality of the product is always desirable on an ongoing basis. Control charts can be established for both variable and attribute data. They can also be established for input or response in a process.

It is recommended that control plans be created first for critical components. The plans should document key process characteristics and requirements, test and data collection methods, and management team composition and structure. The type of control charts to be specified in these control plans is dependent on the data source and data type collected for these critical components. Tables 3.22 and 3.23 provide guidelines and tips for selecting appropriate control charts and improving control chart implementation. The arrows indicate the degree of return on investment, from the least to the most.

Table 3.22 Guidelines to select proper control charts

Data type\Data source	Input	Output
Variable data	Hard to implement but the most informative.	Not easy to implement but more informative and lagging.
Attribute data	Not easy to implement but more leading indicator.	Easy to implement but less informative and lagging.

Table 3.23 Tips to improve control chart implementation

Do's	Don'ts
Select control charts for critical few process variables instead of all/most variables encountered during development.	Implement without training the operators on how to read and react to control charts.
Use software that can provide real-time control charts.	Forget to validate control chart software since it usually acts as a "black-box."
Address out-of-control conditions prior to completing validation.	Forget to create a control plan which includes control chart as elements of the plan.

Process capability

Process capability is an important measure that indicates how capable the manufacturing processes are for a medical device. For the critical variables mentioned in the Control Chart section, process capability can be calculated after establishing that the process is under control or stable. The formula typically used to calculate process capability is:

$$Cpk = \frac{(USL - \bar{X})}{3\sigma}$$

or

$$Cpk = \frac{(\bar{X} - LSL)}{3\sigma}$$

Table 3.24 Tips to improve process capability calculation

Do's	Don'ts
Understand the difference between "short-term" and "long-term" process capabilities before using them.	Calculate process capabilities without ensuring that the underlying distribution of data is "Normal." If the data is not normal use non-normal capability indices.
Focus on calculating capability for characteristics that impact the customer or down-stream processes the most. Use FMEAs to decide on which one of the characteristics must be controlled over the long run.	Calculate process capability values if a manufacturing process is not stable. This must be avoided at any cost and the focus should be on stabilizing the process.
Make sure there are sufficient data points during process validation to calculate capability indices.	Forget to establish requirements or baseline for capability prior to validation closure. This will ensure that ongoing production can maintain the capability of processes.

where USL and LSL are upper and lower specification limits for the characteristic that is controlled and σ is the standard deviation. Please note that another measure, Cp, should also be calculated along with this measure during development. This will help the product development team to understand how close to the target the process is in addition to how capable the process is. Without going into the details of how to calculate process capability when the data is not normally distributed, we will identify in Table 3.24 the common pitfalls to avoid when calculating capability.

This concludes our overview of the DFSS tools. In the next chapter we will show how FDA's Design Control guidelines and DFSS are linked so that there can be one integrated approach to implementing an effective Design Control process for medical devices.

chapter four

Design Control and Six
Sigma roadmap linkages

This chapter has the purpose of linking Design Control requirements with Six Sigma. In specific, we will talk about Design for Six Sigma (DFSS) as a business process focused on improving the firm's profitability by enhancing the new product development process. For the most part, if well devised, DFSS will help to ensure compliance with regulations,* though the original aim of Six Sigma programs has always been to positively hit the bottom line and to promote growth. We have chosen the product development domains (PDD) model from Chapter 3 as the DFSS methodology to follow, and the intent is to show that both roadmaps, DFSS and Design Controls, can be walked in parallel and thus take advantage of such synergies. The design control model (Figure 4.1) that we will follow is based on the waterfall model stated by FDA in their March 11, 1997, "Design Control Guidance for Medical Device Manufacturers."** The DFSS methodology is the flow-down requirements/flow-up capabilities mentioned in Chapter 3. Later we will see that we are really talking about classical systems engineering (e.g., requirements management). The waterfall model and the methodology were also discussed in Chapter 4 of our first book. This book introduces the DFSS terms and makes the connection to design controls.

A point to realize from the waterfall model is that in reality, the NPD team is constantly verifying outputs against inputs. So the first myth we are going to mention in this chapter has to do with the false belief that new product development is carried out following a strict set of serial, sequential steps. For example, from Figure 4.1 we notice that design review is really an ongoing process. Though ideal or

* Specifically, design and process controls.
** See www.FDA.gov.

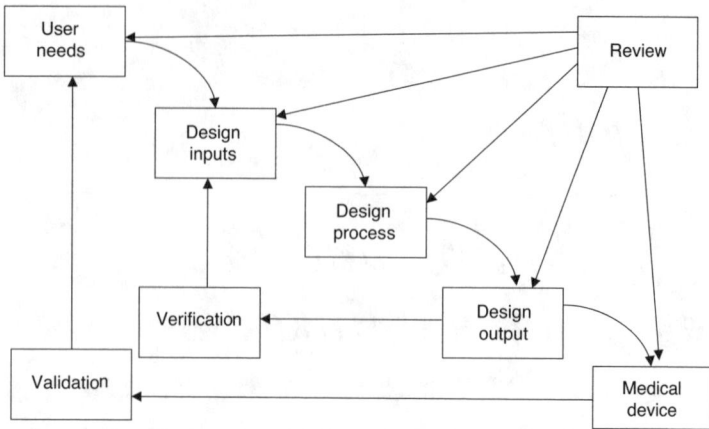

Figure 4.1 Waterfall design process (GHTF).

logical to those who have never designed a technological product, the series approach is neither logical nor optimal unless you are merely copying existing and very well-understood technology and its application. In fact, if the process of NPD is serialized, there is no need for a multidisciplinary or cross-functional team approach or concurrent engineering.

In this chapter, we first start with some background information on DFSS and the medical device industry. The authors believe that it is of utmost importance that those black belts and DFSS leaders coming from other industries understand the state or nature of the medical device industry.

Background on DFSS

What is the motivation to go beyond the DMAIC in Six Sigma?

In times past, black belts (BBs) and quality engineers (QEs) applied statistical engineering methods aiming at uncovering key process inputs or factors that could affect a process. They then used typical quality engineering methods such as multiple linear regression to obtain a prediction model for central tendency and spread (e.g., Taguchi models) and then made predictions about the capability of the process and defined control plans. So far, this is very similar to the DMAIC methodology of the typical Six Sigma program. However, sometimes the process capability or the actual process performance was suboptimal or even inadequate. This led QEs and BBs in manufacturing to find limits to the physics or the science of a given technology, product, or process that inhibited the possibilities of

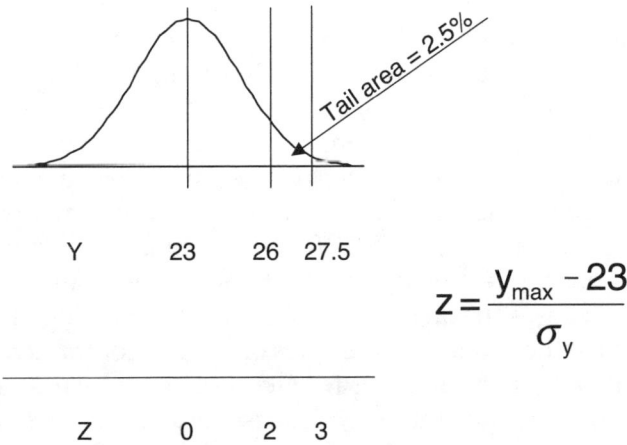

$$z = \frac{y_{max} - 23}{\sigma_y}$$

Figure 4.2 Hypothetical example where an incapable process has been designed for failure.

achieving better than Three Sigma quality levels. These limits had been defined based on Six Sigma methodologies such as DMAIC, employing tools to evaluate process stability (e.g., SPC or other sequential testing) and tools to evaluate potential factors of noise and signal affecting the process (e.g., Taguchi, Classical DOE, or a blend of both). However, it was not enough. Let us see the following example:

$$\text{output} = y = 8 + 3x$$

where x is the setting of a process parameter with a functional discrete range between 5 and 6. If your maximum output limit is specified as $y = 26$ and the process has a natural noise level described by the standard deviation on y such as:

$$\sigma_y = 1.5$$

then, when $x = 5$, the process is centered around 23. At three standard deviations or Three Sigma (23 + 1.5[3] = 27.5) from the center of the process, the probability of producing a defective product is described as $P(y > 26) = (z > 2) = 2.5\%$ (see Figure 4.2).

See that if x is set to the other possible value, 6, the percent defective would be worse than 2.5%. If the physics of the manufacturing process cannot allow the x to be set at less than 5, then the process is not capable by virtue of its own design. There is not much that the manufacturing plant can do other than implementing 100% verification of product.* The manufacturing personnel may be perfectly efficient and accurate following the procedures and documentation (cGMP "perfectos"), but this does not change the fact that the process is incapable and there is very little that factory engineers could do to change this reality. In cases like this, the responsible parties for process development did not produce a manufacturable process or it was not "Designed for Six Sigma." We have also seen the case where the technology was not mature enough to be on the market. This causes the factory personnel to start making unnecessary adjustments to the process, sometimes obtaining contradictory results of experimental design leading to overall confusion and chaos. It is important to state that the very first issue faced by many medical device manufacturers is the fact that the relationship between inputs and outputs is unknown. That is, the manufacturing process flows down from the NPD organizations to manufacturing (e.g., design transfer or knowledge transfer) without prediction equations. In many cases, nobody knows the meaning of the specifications or tolerances. Who can make a connection to functional and to customer requirements?

On the other hand, the job of the quality engineer or black belt is also to question the need for the spec to be a maximum of 26. Typical DFSS/QE questions are:

- Where did the specification come from? What does it mean?
- Is it directly related to a customer requirement? In which way? Is there a relationship** between this process specification and the customer requirements?
- What is the consequence if we ship the product out at 27.5? Who knows? How can anybody know, if traditionally specs are not necessarily justified in the Design History File?

---

* See the process validation guidance from the Global Harmonization Task Force at www.ghtf.org. Also, verification is explained in Chapter 3 of our first book.
** A relationship is ideally described by a mathematical formula. In DFSS we refer to it as the transfer function. This term is new, but the concept is very old. The transfer function is nothing else than a prediction equation. The authors will credit Genichi Taguchi for his concept of parameter design and the spread of multiple linear regression as the analysis tool. Taguchi simplified DOE and its analysis and showed simple ways of implementation, opening it to the world of the non-statisticians. We will also credit the book from Schmidt and Launsby *Understanding Industrial Designed Experiment* with a significant push of the concept in a simple and practical fashion.

Later on we will discuss how the enhanced design history matrix*
can be used as a tool to manage and track design requirements** and
V&V activities during the design and development cycle.

The example above was also aimed at illustrating in a very simple
way why it is said that the Six Sigma DMAIC process is reactive. In
the medical device industry, DMAIC is mainly run by manufacturing
or operations personnel. As typical of this industry, manufacturing
personnel are paid for producing, not for designing or developing.
Therefore, Six Sigma programs in manufacturing have many limitations when the gap between process capability and the customer
requirements (e.g., the maximum tolerance or spec) is wide. This
comment is rooted in the reality of today's industry regarding medical
device manufacturing. The reality is that the engineers in the factory
do not typically have the knowledge, experience, or the time to
"reverse engineer" a device's design and truly understand the impact
of changes or deviations to the device's intended use. The opposite
would be a manufacturing process in which only a few significant
factors have to be controlled and the improvements do not require
major technological changes. A DFSS program should help in filling
in this lack of technical expertise commonly found in medical device
factories.

Background on the medical device industry

The purpose of this section is to briefly and superficially discuss some
of the peculiar issues that the Six Sigma implementers will find in the
medical device industry. This is an industry where companies may
fail to extract value as a driving function of the business because:

- The regulated nature of the business makes it a bureaucratic
  one by default.*** For example, manufacturers in many other
  industries do not have to keep track of so many detailed deviations to procedures, specs, and so on, while in this industry
  these are basic cGMP rules. There has to be control of every
  little step in the factory environment since there always has to

---

* In DFSS terminology, this will be an enhanced design cascade or flow-down across the design
domains. This concept was introduced in Chapters 2 and 3 of our first book. It is also known
as requirements management and requirement cascading.
** Requirements of the design of the device and the manufacturing process including packaging
and sterilization.
*** The authors consider regulation as a necessary evil. By just looking at the list of recalls in
the FDA MAUDE database and Medical Device Reports (MDRs), it is easy to realize why we
need regulatory bodies out there.

be accountability and responsibility well stated and recorded. A typical MDI factory includes "cages" to segregate material and finished product, and it also requires an independent quality control unit that does mostly documentation work. In the MDI, there cannot be such a thing as "empowered" employees who can "correct" the process when they think it is appropriate to do so.

- The practice of medicine is said to be an inexact one. Standardization of medical procedures does not equate to the best health care since medical judgment and practice is still dependent on the health care giver's view of the illness or condition as well as the school of thought to which they belong. This and the litigious society in the United States nurture a large number of liability suits and public pressure on politicians and government agencies. Who is to be blamed for the death of a patient? The drug makers? The doctor? The medical device manufacturers? Or just mother nature?

- Medical procedures do not have an exact "transfer function." That is, there are many unknown variables that can affect clinical outcomes, and many well-documented clinical studies are not reproducible or are contradictory of each other. For example, the monthly specialized newspaper *Gastroenterology & Endoscopy News** says in the September 2003 issue that there are mixed results when comparing research papers on the clinical effectiveness of urgent colonoscopy. So, if you are a medical device manufacturer supplying this market, isn't your Voice of the Customer (VOC) contradictory? Every day more and more colonoscopies are being done.

- There is a time to market and a time to grow acceptable adoption rates. In some cases, lengthy clinical studies may be needed. This may take time that a commercial technology company may never have to wait. In fact, a typical rule of thumb out there for medical device start-ups is to never try an initial public offering (IPO) unless they are profiting from their products. At the same time, the exit plan of many of these little companies is to be purchased by a big one. For all these reasons and others, the level of characterization and "cascading" of requirements is typically very limited for medical devices. The entrepreneur has to Design for Acquisition (DFA) before they burn all the venture capital (VC) they may have.

---

* See www.gastroendonews.com.

- Emerging commercial technologies do not enter the healthcare industry as fast as they enter the mass markets. Specifically, if the technology comes with new approaches or concepts, the NPD project has to consider the learning curves that include the many white papers that the few leading healthcare professionals would write to let the many followers know about the new gadget or new technique out there. This part of product planning may run in parallel with product development but it should ideally be defined up-front. Here is where the DFSS project charter discipline comes in handy.
- Typical company politics. The bigger the size of the company, the slower they are to adopt or develop new technologies and change. This is exasperated by the functional arrangement of the personnel. For example, in some companies the quality systems function may not understand new technologies brought in by the acquisitions group. However, they may want to impose on electronics the same "quality system procedures" that they have been using for mechanical components. In another company, all their devices are mechanical in nature while all their QC personnel have a degree in chemistry.
- How do you benchmark when you are a market leader? Many healthcare and medical device market segments can be described as quasi-oligopolies. Only a few guys compete. For example, Baxter and Abbott Labs dominate the saline solution (IV sets) market. Other than logistics and pricing, what can be done to differentiate the products from each other?
- Many MDI companies make the terrible mistake of focusing on protecting the past rather than building the future. This bad strategy is nurtured by many leaders in regulatory affairs, quality systems, and compliance groups who believe they would look bad at the firm level if admissions of mistakes in the past may jeopardize careers. For example, while training and coaching quality engineers for a major MDI company, one of the authors was given two sets of data. The data was the output of a destructive test aimed at evaluating the most important characteristic of the device. Both sets of data had the same average, but very different standard deviations. As a statistical engineer in the MDI, we know that error of measurement is a big deal in this industry. Upon asking about the measurement technique for both data sets, the quality engineers indicated that the set of data with the large standard deviation had been generated using a manual technique, while

the second data set had been generated using an automatic one. Obviously the problem was the test method at hand. However, the firm decided that they could not switch to the automatic technique because "regulatory and quality said that switching over could imply problems with the product already in the distribution channels."

- Insufficient funding (e.g., small companies may not have all that is needed to present sufficient objective evidence or just enough for adequate assessment of the VOC; besides, a PMA submission will cost more than $200,000 U.S. in 2004*).
- Reliability engineering in the medical device industry that we know today began in the late 1960s. This is 30 to 40 years late when compared with other industries. One important fact about reliability engineering is that the basic mathematical methodologies such as parallel and systems reliability prediction are easy and straightforward to understand by those who know probability and statistics. What is always a mystery is that specific applications that a company may choose to arrive at the estimation of mean life and failure modes. In fact, the reliability methodologies, including life testing, can be as valuable to a company as its intellectual property. A way to overcome this shortcoming is investing long hours and many resources to do modeling and analysis. This is typical of aerospace and semi-conductor industries (i.e., high-technology firms). If we think of the number of recalls in the medical device industry (see later in Chapter 6), and think of an equal number of recalls of commercial airplanes, we would not dare to fly! Table 4.1 depicts a good relationship between aerospace and medical device reliability.

Sometimes, the best medical device solutions might imply a need to wait until the design of the next generation for a product line. However, what if survival depends on positive cash flow for the short term? In the United States, historically, whoever comes to the market first with a breakthrough device is the one that will typically make the most profit. In fact, second-place finishers will typically have to come in through the back door (e.g., either a very compelling reason such as obvious technological improvements or cheaper price). One of the best examples is in the *in vitro* diagnostics (IVD) field. Hospitals or clinical laboratories have to buy or lease a very expensive piece of capital equipment (e.g., the analyzer). It is not only the capital

* Source is the MDUFMA article in *The Silver Sheet* (Vol. 6, No. 12, December 2002).

Table 4.1 Reliability Comparison Between Aerospace and Medical Device Industries

Aerospace equipment reliability	Medical device reliability
Costly equipment.	Relatively less costly equipment.
A well-established reliability field.	A relatively new area for application of reliability principles.
Large manufacturing organizations with an extensive reliability-related experience and well-established reliability engineering departments.	Relatively small manufacturing organizations with less reliability-related experience and less-established reliability engineering departments.
Reliability professionals with extensive related past experience who use sophisticated reliability approaches.	Reliability professionals with relatively less related past experience who employ simple reliability methods.
Well-being of humans is a factor directly or indirectly.	Lives or well-being of patients are involved.

Adapted from B.S. Dillon, 2000, *Medical Device Reliability and Associated Areas*, CRC Press.

investment, but like any other computerized equipment, users need specialized training and, of course, go through the learning curve. This is another up-front investment the user has to make, and yet they have not started to do any business.

Therefore, whichever medical device company is out there first placing their analyzers will gain a competitive advantage. The second-place contender would have to make exorbitant concessions to convince the customer to dump the previous acquisition and the resources invested. Thus, the race was not necessarily to deliver the "perfect" product but the one with the minimum acceptable customer-wanted features. In summary, DMAIC for Six Sigma may give a close-to-perfect product a little too late. A well-conceived DFSS program based within the proposed road map will help to define the set of minimum customer requirements and to provide the foundation for better long-range planning (e.g., based on multigenerational plans for products).

The road map to achieve control of design "a la DFSS"

Table 4.2 is our "equivalency" table that links Design Control requirements with DFSS. The first column is the flow-down requirements/flow-up capabilities from the DFSS methodology. The second column is the domain in which the NPD process lays. However, Figure 4.3 clearly shows that for any NPD process, the domains overlap with each other, and the innovation domain should have all others embedded in it. In fact, the reality of the MDI is that it relies for the most

Table 4.2 Equivalency table for FDA Design Control requirements and DFSS

DFSS	Domain	Waterfall model and FDA guidance on Design Control	Typical DFSS deliverables or actions in the MDI
Define the opportunity[a]	Innovation	Not applicable	Project charter, business case, and project-related risks.[b]
Customer wants and needs	Customer[c]	Design and Development Plan	Project Plan (same as Design and Development Plan).
			Plans for obtaining VOC.
		User needs or potential needs	Obtain VOC: Raw and analyzed customer data.
		Start DHF	Define what makes the product
		Design review	special.
			Define what makes the potential device hazardous and find out why.
			Translate the VOC into high-level systems requirements.
			Define the performance requirements for the system requirements.
			Initiate requirements management (cascading).
			Initiate project's dictionary (medical specialty jargon).
			Reassessment of medical or clinical risks.
		Design input	Measurable high-level systems
		DHF	requirements and
		Design review[d]	performance.
			High-level systems requirements prioritized and with targets or acceptance ranges.
			What would the future user (e.g., customers) evaluate when presented with the final device?
			Medical Device Report (MDR) analysis (e.g., use the MAUDE database at www.fda.gov).
			Analysis report complaints from similar devices (when applicable).
			Continue requirements cascading.

(continued)

Table 4.2 Equivalency table for FDA Design Control requirements and DFSS
(continued)

DFSS	Domain	Waterfall model and FDA guidance on Design Control	Typical DFSS deliverables or actions in the MDI
			Review healthcare industry publications such as the *Gray* and *Silver Sheets*.
			Add definitions to project dictionary.
			Research healthcare community journals. Evaluate potential sources of harm such as the ERCI.
Medical device functional requirements	Functional	Technical design inputs DHF Design review Update Design and Development Plan	Lower level or tier 2 requirements.
			(Engineering terms) or subsystem requirements.
			Product concepts, prototypes, or computer-modeled devices.
			Continue requirements cascading.
			Add definitions to project dictionary.
			Risk analysis such as the Systems FMEA.
Design and develop medical device, its packaging, and manufacturing process.	Design domain and process domain	Approved design inputs Update Design and Development Plan Design output Risk analysis Design review Design changes DHF	Chosen Best Concept.
			Component requirements.
			FMEA/FTA, functional block diagrams.
			Reliability goals and test plan.
			Manufacturing Process Flow Chart.
			Quality Control and Assurance Plan.
			DFx analysis.
			Component and product parameter optimization.
			Process parameters and operational ranges (characterization).
			Process optimization.
			Add definitions to project dictionary.
			Continue requirements cascading.

(continued)

Table 4.2 Equivalency table for FDA Design Control requirements and DFSS (continued)

DFSS	Domain	Waterfall model and FDA guidance on Design Control	Typical DFSS deliverables or actions in the MDI
Verify and validate product and manufacturing process	Design and process domain	Design outputs Design review Design verification Design transfer Design changes Design transfer Risk management Design review Design validation DHF	Validation of test methods. GR&R, nested DOEs for IVDs. Component reliability testing. Process characterization. Component qualification (including process capability and stability at that level). Subassembly/assembly overstress testing. System integration (e.g., can all the parts work together?). Subassembly process capability and stability. System process capability and stability including sterilization and packaging. Add definitions to project dictionary. Complete requirements cascading.

[a] It is very typical these days to look for alternatives to high selling drugs or expensive treatments.

[b] Typical risks are, for example, not getting approval from the FDA in a PMA or not finding an IRB capable or willing to approve a clinical study on an unknown technology. Another risk is the eventual transformation of medical science to a different treatment or drug or biologic that is not even in the picture now.

[c] Not only those who will pay the bills, but keep in mind that MDI is a regulated industry, thus the FDA, EPA, and DEA, among others, are customers as well.

[d] At this early stage of the process, the review is merely aimed at making sure the potential needs or actual needs are understood and that the process of translating into systems requirements makes sense.

part on existing technologies such as materials, processes, components, and methodologies. This fact combined with the need for speed to market causes the design and development process to be one where early in the process many single requirements can be "instantly" cascaded down by NPD team members who, for example, are experts in materials and processes. In other words, this means that not all VOC may have been gathered. For example, someone on the NPD team may already know that the only biocompatible material to be used in a given design may result in an infeasible project. This is an advantage of concurrent engineering, where early notes like this can

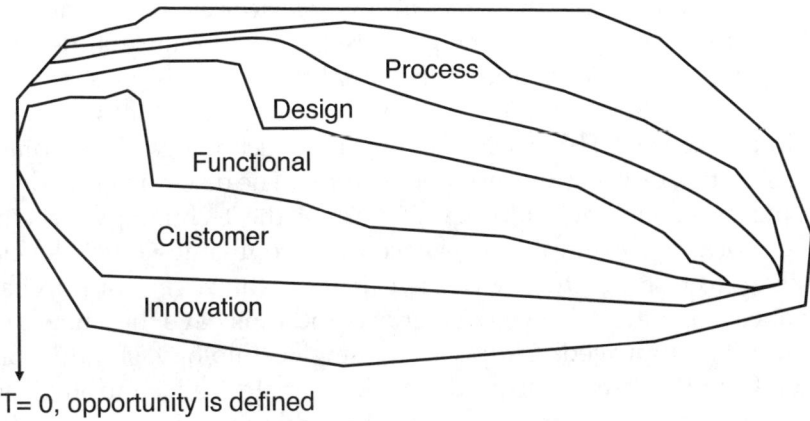

T= 0, opportunity is defined

Figure 4.3 Domain overlapping.

save time and resources. This would mean that the MDI has to become more dynamic and willing to accept and live in a constant state of change. See in Figure 4.3 that all domains may start showing up some area of application close to the genesis of the project or T = 0.

Pre-requirements to DFSS and Design Control (innovation domain)

Both DFSS and the application of design controls assume that the "research" part of the R&D work, or the "development of technology and science," has been done and that the technologies involved in the potential concepts have shown to be feasible (e.g., robust, measurable, controllable, predictable, understandable, and the clinical effect reproducible). In other words, there is no technology evaluation to be done, just the specific application. A typical limitation with new technologies is the unknown aspect, only now the industry is beginning to recognize that "technology development" projects should be largely independent from "product commercialization" projects.* When a project is deemed to be commercialized, expectations from management exert such pressure on NPD team members that it may influence team members to "cut corners," thereby resulting in lack of coverage of technical sensitivities in the product. A real-life example of the MDI and immature technology is IVD, in specific, immunoassays. Those with experience in this trade know that the number of recalls and complaints are always higher than any other medical

* Creveling, Slutsky, and Antis, *DFSS*, Prentice Hall PTR.

device. They also say that it is impossible to develop an accurate transfer function that could help to predict performance; thus, manufacturing of polyclonal and monoclonal antibodies is considered an art. However, the technology has been out there for at least 18 years.

In this sense, DFSS discipline forces the technology development teams to stick to a very important discipline. The discipline is to define the measurement methodology as part of the technology development work. For example, pharmaceutical companies must develop an assay to test the efficacy of their drugs if they ever want to have a chance to get their drug characterized and thus have their new drug application approved. Simply put, new technology should come down from the researchers with a well-characterized, controlled, and standardized measurement system to be able to validate the design. We are not talking about GR&R or nested models for error of measurement estimation only. We are really talking about the meaning of the measurement. What does it really say? How do you know? In this sense, we can see how the DFSS program will help companies to be in compliance while they pursue business profitability.

Define the opportunity (innovation and customer domain)

Patient, healthcare giver needs

The first thing to keep in mind in the MDI is that most medical devices can be classified as tools, aid mechanisms, monitoring systems, and automatic systems for diagnosis and treatment. Thus, in principle, some medical device manufacturers have a lot in common with manufacturers such as Black and Decker, Makita, and Craftsman. Just as the tool maker has to know how to use the tool, the makers of monitoring and aid mechanisms have to know the medical sciences involved. Identifying customer needs requires out-of-the-box thinking or being in the innovation domain. In reality, identifying customer needs requires the mind set of the tool user. Forty years ago, a good scalpel (e.g., sharp and rigid) would have been enough to a surgeon. How did anybody decide the length of the blade? How sharp is sharp enough? At that time, asking about ergonomics in a minimally invasive tool (e.g., laparoscopy) would have been a waste of time. The procedures were unknown and it was not until endoscopes (e.g., new technology) became available to healthcare providers that innovation led to minimally invasive surgery. The authors have seen this kind of reaction upon presenting a surgeon with a new concept: "Why are you designing a device (e.g., tool) that I do not need?"

Gathering VOC in this industry requires understanding of the specific medical practice so that the questions (e.g., surveys) can intelligently assist the innovation process. In fact, in many aspects, innovation would be required to anticipate or speculate the future of medical practice. In many cases it can even imply redefining future medical practice. Therefore, statistically justified customer surveys or interviews may not provide the required input for design. VOC at the innovation stage is speculative or "over-the-horizon." VOC is partially useful to define what is needed. This is why our past experience tells us that the first contacts with potential customers should be based on open-ended questions. Why? Because statistical methods were invented to draw conclusions from static populations. Innovation implies the creation of such a population.

The medical device field is a dynamic population crowded with all kinds of uncertainties; and uncertainties bring risk to the business based on both regulatory and technical uncertainties, among others. The technical uncertainties are comprised of clinical effectiveness, acceptance by the medical community, clinical cost-effectiveness, and manufacturing cost-effectiveness. Instead of using "customer data" to interpolate (e.g., draw a conclusion based on the sample), the data will be used to extrapolate (e.g., establish a hypothesis and set up an experimental plan). In many cases, all that is needed is the attention of a leading healthcare giver who, once the device is developed, can influence others in the specialty field. This is like trying to sell a new brand of sports shoes. As depicted in Table 4.2, the initial stage of the innovation domain, or defining the opportunity, is purely a business activity that proposes to a firm a potential chance to define a value proposition to the healthcare community or patients. Therefore, according to the 1996 preamble to the Quality System Regulation (QSR), these activities are not to be regulated by the FDA.

We believe that the spread of the Six Sigma concept across all businesses in the United States and abroad, and among big companies and management consultants, may define the typical project charter as the standard business proposition of this decade. Therefore, whether you are part of a large firm or just an entrepreneur, it would be worthwhile to learn what a project charter "a la DFSS" is all about.

About the customer in the medical device industry (customer domain)

We reiterate that the business opportunity team should know what the healthcare professional knows. Keep in mind the three typical elements of the value proposition on medical devices are risk/benefit ratio, safety, and effectiveness (e.g., intended use). The first thing we are advising is in assessing the variability among the professionals involved. Assess their beliefs, whether accurate or not, and more importantly, their perceptions. For example, gastroenterologists who perform endoscopic procedures ("endoscopists") deal with certain patients who develop a narrowing of their esophagus called strictures. These conditions require a mechanistic procedure called esophageal dilation (reopening of the tube). In this field of healthcare giving, there is one school of thought that believes that the time stretching the esophageal tissue is not relevant at all while the other school of thought says that a minimum of 60 seconds stretching the esophageal walls is important. Which doctor is right? Does it matter? Are there any white papers comparing the two hypotheses?

In fact, knowing the existence of both is relevant to the R&D or NPD Quality and Reliability Engineer who will define the reliability requirements and the testing strategy. In a case like this, the device will be defined with a mission profile assuming the worst case (e.g., $2 \times 60 = 120$ seconds stretching the esophageal tissue, which is a safety factor of 2). Understanding the customer wants and needs is called VOC in the DFSS jargon. Our message here is to reiterate that the NPD team must understand the medical procedure(s) and go beyond what a few potential customers say. Also, reliability planning is totally dependent on VOC, and it starts this early in the NPD process.

Those in charge of VOC should assess the different schools of thought among healthcare professionals, as this may be the foundation to define the market segmentation or to define a "robust" marketing strategy. Understanding variability among the customer groups will avoid overreactions to contradictory customer feedback. It is this overreaction that causes teams to lose focus and become diluted in trying to meet the needs, nice-to-haves, and specific requests from all potential market segments. In-depth customer knowledge as well as the drivers that define the different segments must be explained in detail to the NPD team. The fallacy that only marketing or business development groups are supposed to know and be accountable for the customer domain inputs is not in line with DFSS principles.

As a reminder, the sponsor teams must ensure that the detailed set of customer requirements is fully understood by all team members. A good start is to ensure that the anatomy and clinical procedures are also fully understood. We want to emphasize this since it is so important in the medical device field. The DFSS sponsor teams (e.g., senior management sponsoring an NPD project) must ensure that all team members have a minimum level of clinical knowledge that goes beyond the typical anatomy or biochemistry lessons.

The project charter

As inputs to the design and development plan, the project charter has defined the objectives and elements of the scope of the project. Typically a function such as business development or strategic marketing is the driver and main participant of the data gathering to creating the project charter. Before design controls kick in, questions like the following are already answered or on their way to being answered:

- What market segment(s) are we targeting (e.g., outpatient clinics vs. hospital*)?
- Are we providing a device to answer an existing need or are we leading a new medical procedure (e.g., the launch of the device requires specialized training to healthcare givers, white papers written by the specialty leaders, special regulatory considerations, clinical and technical risks)?
- When does the device need to go into the market? Is this a global product (e.g., compatibility issues with European operating rooms vs. domestic)?
- Do we want to claim compliance to a voluntary standard? What happens if we don't (e.g., regulatory strategies and submissions)?
- What is the pricing strategy (e.g., this will define the Cost of Goods Sold [COGS] to be met as per company internal goals for a good return on investment or NPV)? What are the margins and the anticipated volumes?
- The medical device manufacturer has to see the nature of the product and technology as part of the value proposition. For example, what is the value proposition when capital equipment designs are compared against single-use device designs? Also, servicing may be as important as performance when the equipment is brand new.
- What are the risks of this project?
  - Technological (e.g., first time the technology is used).

---

* As health care is affected by technological changes, so are the potential segments. For example, as more minimally invasive procedures are done, the purchasing function of hospitals will decrease, while outpatient specialized suites (e.g., endoscopic centers) will increase.

- Regulatory (e.g., this is a first to market, no predicate device, may need to generate clinical data).
- Business (e.g., what if a competitor comes up with a better concept?).
- Manufacturing (e.g., what if the factory cannot handle the new technology?).
- Financial (e.g., the capital requisition order may sound too risky).

See that a full cross-functional team may not be needed to find adequate answers to these questions. For example, you may not need your day-to-day manufacturing engineers yet, but you will need a manufacturing development engineer who can possibly generate a strategy to use current facilities for the new product. On the other hand, some design engineering folks may be of great help to business development or marketing in the evaluation of technological concepts.

We have basically defined the role of the project charter as a DFSS pre-requirement to the design and development plan. We want to clarify that they are not the same. The project charter is a formal business document of no relevancy to the regulatory authorities. A design and development plan is a formal auditable document required per FDA's QSR and ISO 13485 standards.

Customer wants and needs — design inputs (customer domain)

As part of the next steps after chartering the project, formal and more comprehensive data gathering from the potential customers is planned. Thus, VOC gathering shall be seen as an iterative process, just like product design and development or problem solving. What we see in practical application is a process of fine-tuning the processed data until a set of refined statements can be made about what is needed to be designed (e.g., intended use). Figure 4.4 depicts this iterative process.

The design and development team needs continuous access to the customer. This access, in the form of interaction, should be well planned. The DFSS tools mentioned in Chapter 3 will help the design and development team to increase effectiveness in this data gathering process. Design input is not only the physical characteristics of the medical device, but the entire product such as storage requirements, documentation, labeling, servicing, parts, components, accessories, and so.

From Figure 4.4, at T = 0, many open-ended questions were raised like in any other investigation. Different customer clusters or strata

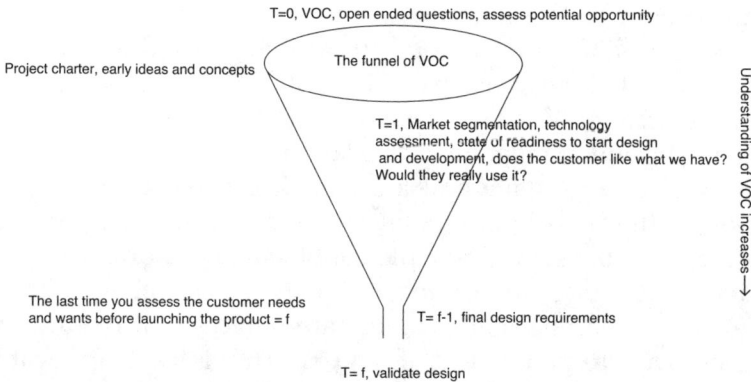

Figure 4.4 VOC for medical devices, new products, or new technology.

were included in the data gathering. A design plan fed by the project charter is then generated. Typically, this plan includes more VOC to be gathered (e.g., to refine the data set). At T = 1, the business strategy has taken shape and the project charter had defined the segments to be targeted. A second visit to the customer base is necessary. The design and development team may now have a hypothesis, ideas, and maybe early prototypes (e.g., concepts) to show to the customer. The questions at T = 1 will be more targeted at finding specific answers, and to set more accurate and specific direction. Existing design inputs are then further refined. Other visits are typically expected. Companies cannot pretend to gather all the VOC needed in a big giant study at T = 0. Remember, during the early stages you do not know what you do not know. In our experience, most healthcare professionals need to see and feel in order to provide any useful feedback.

At T = 1, the team should go to some of the original customers (e.g., your point of reference or experimental control) interviewed at T = 0 to corroborate the already existing inferences and to have early ideas or concepts evaluated. By going to other potential customers at T = 1 you might be increasing your sample size but this may not be the best strategy, because you might be "confounding the experiment" since you do not have a point of reference. For example, it sounds logical that a 55-year-old surgeon would resist adopting a new surgical device if he has been using the "old" gadget successfully for the last 25 years. Thus, gathering VOC with the old doctor at T = 0 and later, when you have a working prototype, showing it to a young surgeon may confound the VOC analysis. It is possible that the 55-year-old surgeon may never be willing to change.

Our recommendation is to always keep at least one point of reference to normalize your second, third, and posterior verifications with the potential customer base. Each iteration during design and development should grow in terms of meeting more customer requirements, and to meet them to a larger extent. In summary, DFSS brings tools for this stage and a set of disciplines to adopt. Those engineers with good experimental design disciplines will be helpful in this stage, as the same principles of blocking, randomization, and controls are of utmost importance to the NPD project.

We want to emphasize that there are devices that require prototyping in order to generate useful VOC. This is very applicable to those devices that fall in the "tools" category. Most physicians, if not all, need to feel, see, and touch the tool or device in order to give a rational opinion and assertively state their likes and dislikes. Showing a 3-D computer-generated model will not be good enough. It becomes a chicken-and-egg dilemma. Should I gather the VOC first? The answer is yes, before any prototypes are generated, and yes, after prototypes are available. Some authors call this design input phase "Concept Engineering." The T = 0 data gathering should not be aimed at being a final study but just the opposite. Open-ended questions are the key. The VOC researchers shall know as much as possible about the anatomy and human science involved; however, they should assume that they do not know what they need to know.

The following is a list of tips to keep in mind while working in the customer domain:

- Ensure access to the customer at all times during the product design cycle for every team member (not just the marketing department).
- Avoid confounding of data, especially when you have no idea of the segmentation of the market. Keep in mind the concept of experimental error and why you always need to reproduce the measurement.
- The first and most important resource requirement is knowledge. Specifically, knowledge about the human body (e.g., anatomy, hematology, immunology), the medical treatment, the therapeutic path, the scientific principles involved in the predicate device(s), and its interaction with the patient and the user(s).
- Do not lead the customer that is surveyed, let them lead you. However, be aware of the unspoken word. There is typically a lot more than what the customers are saying.

- The importance of documentation is not only the evidence the firm needs to show FDA inspectors and ISO assessors that they have compiled design inputs. It becomes the "norm" or standard by which many decisions will be made later in the design and development process. Defaulting to the VOC is the easiest way to solve the typical design project disputes that should normally be expected of team members. In fact, this is why the QSR requires a way for solving conflicting requirements.
- DFSS is said to be a program under which cross-functional teamwork is facilitated (e.g., concurrent engineering). However, in the medical device industry, if the medical or anatomic subject is not mastered by the team members and their design reviewers, the cross-functional aspect may never bring the expected returns. For example, reviewers from manufacturing, demand management, quality sciences, quality systems, regulatory compliance, and "Six Sigma black belt" groups will prove to be mostly bureaucratic obstacles during design reviews if these elements of knowledge are not part of their skills set.
- Play with the predicate device. Reverse engineer, use it out of the recommended conditions, assess reliability, and find weaknesses of manufacturing or design.

All data gathered can also serve the strategic purpose of product family or multigeneration product planning. Thus, VOC is the foundation for design inputs and for strategic planning as well.

Survey and interviews are the default tools for marketing research. The most relevant pre-requirements before you start surveying customers are:

- To understand the human anatomy (e.g., human and animal, biochemistry for IVDs).
- To understand medical practice (whether your device is a technological breakthrough or just another catheter).
  - To understand their jargon. For example, for cardiac surgeons, IMA stands for "Internal Mammary Artery," while for academic anatomists and general surgeons, IMA stands for "Inferior Mesenteric Artery."
  - To understand the current mind sets and limitations. For example, today there are many hospitals facing cash flow or profitability issues. Expensive capital equipment may need some financial assistance to help customers be able to afford such large investments.

- To understand health insurance reimbursement policies. For example, most family physicians and most internal medicine doctors know that the specialists to whom they refer their patients are paid more for the procedures that they perform on patients than for "seeing the patients." Similarly, some specialists know that the general surgeons have a higher incoming cash flow because surgeries are procedures as well. Your new medical device idea may not be appealing to general surgeons, but it may appeal a lot to the specialist if your innovative idea implies a procedure that can be performed in the office.
- Note that VOC is already being heard during the innovation domain, implying that there is no such thing as the perfect execution of steps in series.

Table 4.3 depicts some examples of typical comments that can be expected from customers in the MDI while gathering VOC. Note that similar to the auto industry as well as to personal computers, it is not only the device that matters, but the associated services and human contacts. In modern business terminology, we might say the customer cares about the total experience.

Risk analysis

Risk analysis should be started as early as possible, even in the innovation phase. In fact, for some devices, effective risk mitigation is the actual value proposition. For example, according to a corporate press release (February 23, 2004, at www.BD.com) from Becton and Dickinson and Company, "the BD.id™ System is the first to fully integrate bar-coding technology with proven specimen collection process standards. The system enhances patient safety by helping to reduce the potential for errors during the specimen management process. Specifically, the system helps to ensure that blood and other samples are collected from the right patient, are placed in the proper container, and are labeled correctly."

Management of design requirements (from initial design inputs to final design outputs)

Management of requirements sounds similar to what some defense and high-technology companies had been doing since the early 1980s, especially with the upcoming of relational databases such as those from Oracle and Microsoft. There is a need to define what is needed

Table 4.3 Typical Voices of the Customer in the medical device industry

VOC	Patient	Doctor	Nurse	Biomedical engineer
VOC 1	Clinical effect, minimum pain, no side effects, covered by medical insurance, safe.	Efficiency, easy to use, minimally invasive, better procedure reliability, cheap, with reimbursement codes already assigned.	Easy to use, easy to understand, easy to identify, easy to store, easy to unwrap, easy to dispose, easy to clean, easy to sterilize.	Easy to maintain, easy to repair, easy to diagnose, easy to find and get affordable spare parts.
VOC 2	Easy to use, friendly instructions with plenty of pictures, friendly and knowledgeable voice at the customer help hot line.	Easy to use, friendly instructions with plenty of pictures. Intuitive, so little or no reading is necessary.	Easy to use, friendly instructions with plenty of pictures. Ergonomic and intuitive.	Easy to use, friendly instructions with plenty of pictures and schematic diagrams.
VOC 3	Easy to know when maintenance is needed or something is not working.	With at least one or two white papers published by leading physicians in the field in order to try.	Easy to know when maintenance is needed or something is not working.	With a CD ROM and good technical customer service.

before delivering results (e.g., in accord with the Design Control regulation, inputs come before outputs). This principle is embedded in the QSR Design Control requirements, ISO 9001, and ISO 13485. R&D leaders and DFSS black belts must avoid the typical R&D inventor tendency of designing first, then seeing if a prototype works to eventually do "reverse engineering." We will call this the "shotgun" approach. This approach works sometimes, especially for small start-ups with a great idea and a handful of team members. This inventor has jumped the DMA steps in DMADV and may have just superficially touched the innovation, customer, and functional domains. This is troublesome, especially in cross-functional teams where non-designers will have to sit and wait until the "after the fact" requirement documents are generated (e.g., design inputs and specifications). To the contrary, a formal set of customer and design requirements can become a common target to all team members regardless of the specific skills that they bring to the project. The way DFSS cascades down requirements is depicted in Table 4.4. This table will make sense to a systems engineer* who has been educated in the disciplines of requesting proposals (RFQs), proposal creation, proposal evaluation, systems integration and testing, and systems commissioning. In general, we can say that many electrical engineers and, certainly, software and computer engineers are formally trained to operate this way. The purpose of the last statements is really to create consciousness among management and DFSS leaders to understand the differences among the technical backgrounds. Requirements management is related but not equal to Critical Parameter Management.**

A systems engineering approach to management of requirements

In the mid-80s, high-technology companies like AT&T Bell Laboratories came with companywide design excellence initiatives such as systems engineering and the creation and adoption of a program called DFx (which stands for "Design for x").

The systems engineering role at Bell Labs was fascinating in all the senses of the word. A truly diverse group of engineers with different educational backgrounds*** worked together with the goal of defining the high-level system requirements that could define a telecommunication system capable of meeting future needs and trends of the customer. The key thing here is that this group of systems engineers did not know much about electricity, radio frequency, switching, transmission, and so on, but were very good at understanding customers and in using logic to define their needs or future wants (VOC or design inputs). A second group of systems engineers would then take those requirements and high-level system architectures, and would try to define how to meet the needs with current technology and what new technology was needed to com-

---

* We want to define a systems engineer as that professional, engineer by degree or not, who defines, specifies, and designs the architecture of a system that is to be realized with multiple disciplines such as electronics, mechanics, software, chemistry, and biology. Needless to say, we want to avoid confusion with this term because in the MDI and in pharmaceuticals, systems engineering is typically associated with computer systems only. This is a very narrow view and a typical reason why the function is not seen as needed.

** Creveling, Slutsky, and Antis, 2003, *Design for Six Sigma In Technology and Product Development*, Prentice Hall PTR.

*** One of the authors shared responsibilities with chemists, psychologists, marketing managers, mathematicians, all kinds of engineers, astrophysicists, and operations researchers.

plete the set of customer needs (low-level system requirements or more specific technical inputs). The key element back then and in today's DFSS is this management of requirements; in other words, how the requirements are flowed down or cascaded during the design and development process. The availability of relational databases and other information technology tools will also help to manage the configuration of complex medical systems during design and development. As we mentioned in our first book, design changes occur throughout the entire design and development project. At a given point in time, how can anybody know if the elements of design that had been already approved are affected or not by a change in the components? Who can answer, where did the tolerance or the specification come from, if there have been 50 engineers in a project? Who is the "spec creator?"

The other initiative at Bell Labs was DFx. This implied the need to consider additional goals aimed at the internal company customers to any given "state-of-the-art" design initiative. For example:

- Design for manufacturability
- Design for reliability
- Design for maintainability
- Design for testability
- Design for simplicity

These design goals became an integral part of a successful product design. It was not about making it work, but how to systematically get there with dozens of highly technical and business people who are very specialized in their areas of expertise. These ideas were also combined with Taguchi's concepts of "robust design," the loss function, and parameter design (transfer function).

For the MDI, the evolution started when the bureaucratic and project "watchdog" functional QA engineers and managers became integral to the cross-functional product development teams and were asked to understand the design and development of the product and the manufacturing process. Now these groups had a narrower but more important and relevant role as well as accountability for project results. The next evolution was adding regulatory knowledge so that they could handle the inspector from the Office of Regulatory Affairs (ORA) at FDA. As medical devices become more complex and companies are forced to be more "lean" to improve productivity, these engineers are also becoming systems integrators and systems engineers. Now with DFSS, these groups should evolve into black belts

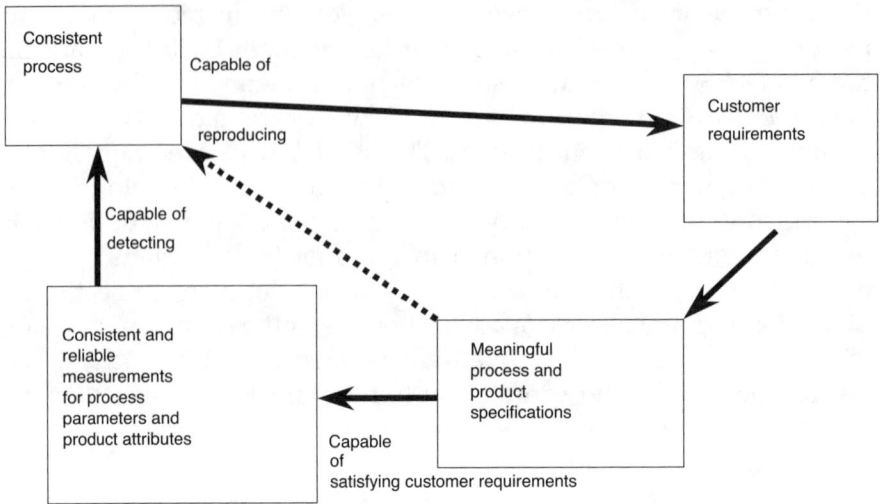

Figure 4.5 Practical view of the business needs.

or master black belts in R&D by adding more business knowledge to their skill set. In medical electronics, there will be systems engineers in the design and development team more involved in the functionality aspects of the device. Whatever the name of the position is, the need of the business is simply having well-understood customer, functional, design, and process requirements* that are integrated into the design and also well documented. If these needs of the business are met and the right disciplines are in place, the firm should have no issues with meeting the Design Control requirements of the QSR.

Initiating the requirements cascade (from the customer domain to the functional domain)

The next evolution for DFSS has been in making sure that the design goals will ensure success in the marketplace; that is, that the right "readings" are taken from the customers or potential customers and that the right translation to system requirements is done. The adequacy of the translation process is in fact very explicit in the Quality System Regulation. We introduced this concept in Chapter 2 of our first book. Figure 4.5 depicts a practical view of the business needs. In DFSS and systems engineering, this relationship is represented by the flow down and up arrows. It is imperative then to realize that design input is a chain of requirements that may require unfolding

---

* Not only the lead designer or inventor or developer but the knowledge has been effectively transferred.

Customer (User Needs)	High Level System Requirements
Shorter hospitalization	From 2 nights to outpatient
Faster healing	From 6 post-treatment visits to 2
Minimally invasive therapy	No percutaneous Incision
Minimum side effects	No abdominal pain
Minimum post-surgery complications	No need for pain killers
Higher clinical diagnostics resolution, precision and sensitivity	Capable of measuring to the nearest mIU, with a CV<10% and a sensitivity of 4%.

Opportunity ▶

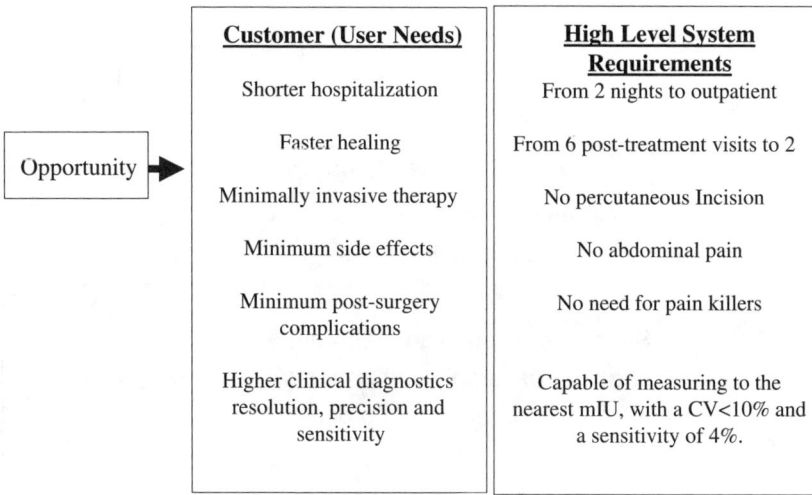

Figure 4.6 Design inputs: Typical user needs in today's health care.

or cascading many times. Therefore, the outputs from the first iteration become inputs to the next. The translation of customer requirements into functional requirements may take several iterations, and keeping track of such activities may be of benefit to the NPD team. This is in fact very explicit in the Design Control guidance from the FDA and the GHTF, specifically when reading the description of the waterfall model. Figure 4.6 depicts a first layer of translation. Note that the high-level system requirements here do not give any idea about how the final design will look. However, explicit measurable output, from a set of design parameters, has been defined as criteria to satisfy the VOC.

What is cascading?

According to the *Encarta® World English Dictionary*, a cascade is a succession of things, e.g., chemical reactions or elements in an electrical circuit, each of which activates, affects, or determines the next.

In terms of requirements management, using Figure 4.7, the cascading starts with user needs at the top of the pyramid and ends with the design details at the bottom. The shape of the pyramid suggests that there will be many detailed design characteristics or parameters for each customer need. The base of the pyramid also indicates that the most fundamental principle of safe and effective design of medical devices is the engineering and scientific expertise. In our experience, developing a dependable and self-directed engineer or scientist in the MDI takes many years, maybe even a decade.

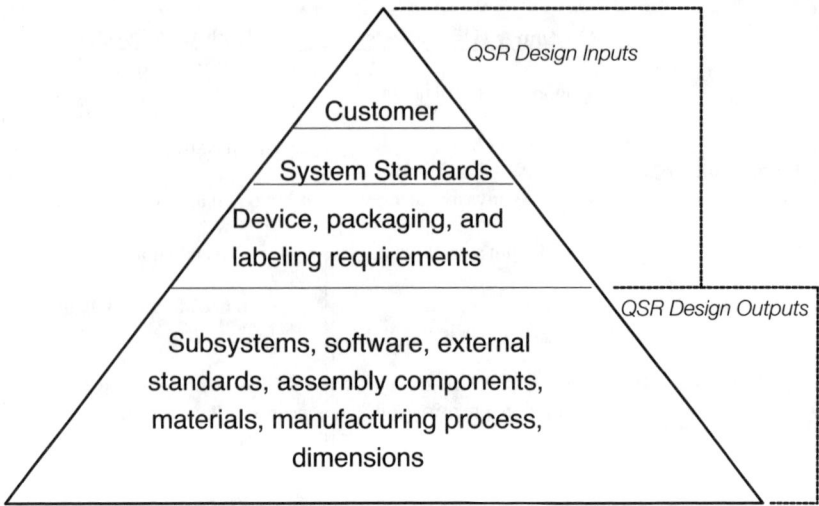

Figure 4.7 Requirements cascade pyramid.

What is the difference between a requirement and a parameter?

A requirement is something that is needed for a particular purpose, according to the *Encarta® Dictionary*. According to *Webster's Dictionary*, a requirement is something essential to the existence or occurrence of something else. This latest definition will be very helpful to realize that the cascading of requirements into lower levels is what the "something else" means. For example, an immunoassay requires a sensitivity of 5%, and the only way to obtain such sensitivity is by designing an assay with a very steep calibration curve at the lowest concentrations of the assay. Thus the calibration curve (a design parameter or variable*) becomes a design parameter from which the functional requirement "sensitivity" depends. In Figure 4.8, we have a typical calibration curve for a non-competitive immunoassay. Sensitivity is typically assessed by exposing the reagents and the instrument to a blank sample (e.g., calibrator A in Figure 4.8B, which is a substance without the analyte that the assay is supposed to detect). If there is no analyte, then there is supposed to be zero reaction. But there is always something detected by the instrumentation, which electrochemists would call "noise" or "background." Thus, the average level of detection plus 2 standard deviations** is inverse regressed (e.g., similar to the propagation of error in Figure 4.8C) into the calibration curve as a concentration and then converted to a percentage.

---

* Different levels of "steepness" will define different levels of sensitivity.
** As defined by the National Committee for Clinical Laboratory Standards (NCCLS).

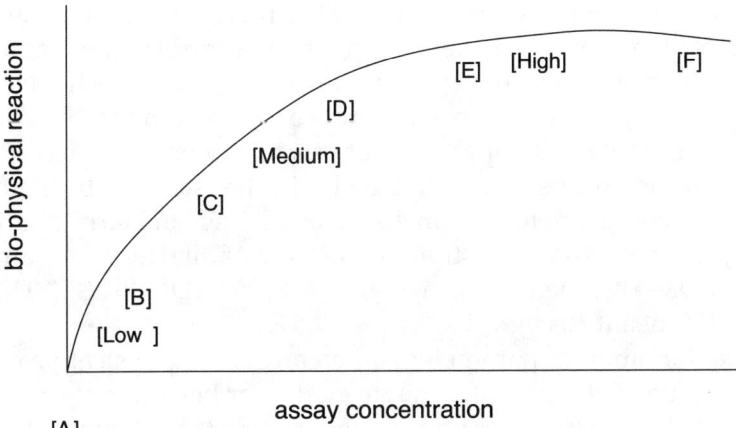

(A)

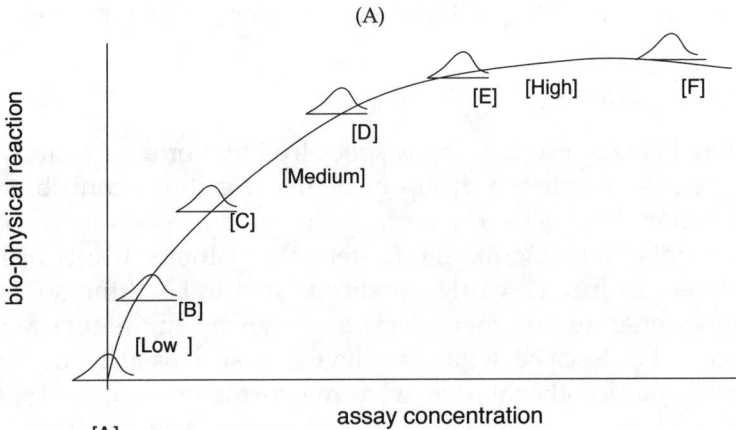

(B)

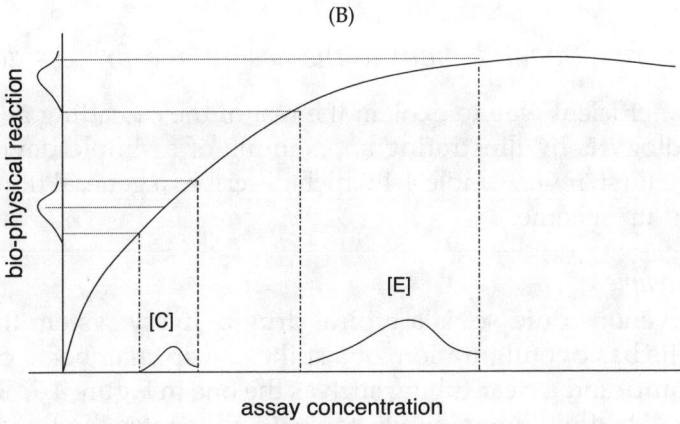

(C)

Figure 4.8 (A) Typical calibration curve for immunoassay. (B) Long-run behavior of targeted concentration. (C)Long-run behavior of targeted concentration.

Interestingly, it is our experience that the mathematics and statistical elements of immunoassay development and control in the diagnostics industry are typically not well understood by the biochemist who designs the assay. It has been in those DFSS companies, or DFSS groups within large companies, that this reality has been recognized, and scientific groups (e.g., personnel with life sciences background) have been complemented with engineers. In fact, in Harry's* original Six Sigma breakthrough strategy, instead of the five fundamental steps of DMAIC, there were eight steps, as in "RDMAICSI." The first letter "R" meant Recognize.

The definition of parameter is then any set of physical properties whose values determine the characteristics or behavior of something (*Webster's Dictionary*); or a limiting factor, a fact, or circumstance that restricts how something is done or can be done (*Encarta® Dictionary*).

What is functional?

Functional is when something is specially fitted or used or for which a thing exists or when a group of related functions contribute to a larger action (*Webster's Dictionary*). So, in the functional domain, requirements that describe the (potential) customer wants or needs are defined. In Juran's world, quality is said to be "fitness for use." The functional requirements, also known as high-level systems requirements, describe what the device is supposed to do for the customer(s) or in other words, what makes the medical device fit for use.

From the functional domain to the design and process domains

The most efficient way to explain the rest of the cascading and DFSS methodology is by illustrating an example of a simple device such an IV set. First, review Table 4.4 which describes a general flow-down and flow-up scheme.

IV set example
An intravenous fluid set is a typical drug delivery system that consists, in its basic configuration, of a spike, a drip chamber, a clamp, a roller clamp, and a clear tubing such as the one in Figure 4.9. Table 4.5 illustrates the flow down, while Table 4.6 illustrates the flow up.

---

* Harry, Mikel, and Schroeder, Richard, 2000, *Six Sigma, The Breakthrough Management Strategy*, Doubleday, NY.

Table 4.4 General view on requirement cascading, flow-down and flow-up, and their relationship to Design Control terminology

Design inputs, flow-down	(Potential) customer wants and needs	Practical interpretation	Design outputs,[a] flow-up[b]	Design outputs, flow-up
↓ (flow-down)	High-level system requirements	How should the product look in order to satisfy customer wants and needs? What other non-unique or old requirements are needed?	Systems assembly, final configuration, and final acceptance specifications	↑ (flow-up)
	Subsystem or subassembly functional requirements (technical design inputs)	What engineering or technical schemes are needed to make the product?	Subassembly, subsystems assembly and testing specifications	
	Component design requirements	What components are needed? What is the role of the component regarding function of the device? What physical, chemical, or biological characteristics are needed of this component? What is the allocated reliability to this component? Why?	Component acceptance specifications	
	Production requirements	How can we manufacture? What technology is needed to satisfy quality, cost, and service level?	Component manufacturing and process control specifications	

[a] All specifications in this column are basic elements of the DMR (Device Master Record).

[b] It is very important to factory engineers and quality engineers to see the relationship between technical specifications or tolerances and the connection to customer and functional needs. What if a manufacturing engineer who had profound knowledge about the customer and the functional aspects of the device is no longer with the company? In many cases, young and inexperienced manufacturing engineers have no idea of the implications behind changes to the fundamental design of a device (e.g., materials, shape, components, specifications).

- Kink-resistant tubing
- Sterile
- Non-toxic
- Pyrogen free

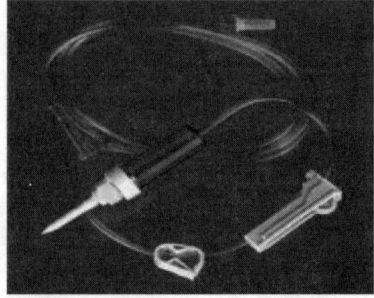

Figure 4.9 Typical IV set.

Transfer function, from high-level systems requirements to lower level subassembly or component design requirements are illustrated. The extra strength of the IV set depends on the bond strength. The bond strength depends on the raw material for the luer lock and on the type of solvent used to create the bond.

$$XS = \text{Extra strong IV set}$$

$$XS = f(BS) \text{ and}$$

$$BS = f(RMLL, SY)$$

where BS is bond strength, RMLL = raw material for the luer lock, SY = type of solvent Y.

Table 4.5 Flow-down cascading for extra-strong IV set

Domain mapping	Requirements cascading	Flow down
(Potential) Customer wants and needs	A suitable IV set is needed for ambulances and trauma rooms.	
High-level system requirements – functional requirement	An extra-strong IV set should have the same functionality, capabilities, and dimensions that a normal IV set has, but with a special strength and a special labeling indicating it to be extra strong.	Extra strong is the functional requirement.
Subsystem or subassembly functional requirements (technical design inputs – design parameters)	The bond strength between a luer lock and the tubing should withstand P pounds of axial force without detaching from the tubing. A 99% reliability level at 95% confidence for a safety factor of 3 is required.	Bond strength is the main design parameter.*
Component design requirements (the raw material and the solvent are a response to the bond strength; thus, they are a design output aimed at defining the bond strength)	The raw material for the luer lock will be X and the solvent Y.	Raw material and the kind of solvent are lower level design parameters.
Production requirements (design for manufacturability)	The luer lock will be made by plastic injection molding controlling the following process parameters: screw speed, barrel temperature, and mold temperature.	Screw speed, barrel temperature, and mold temperature are the process parameters that dictate the output.

* "Parameter" implies variable, something that by changing its value it changes the outcome. In this case, the outcome is the design of the extra strong IV set; in specific, the extra strong capability.

Table 4.6 Flow-up cascading for extra-strong IV set

Domain mapping	Performance requirements	Flow up (process and design capabilities)
Component manufacturing and process control specifications (design verification, design transfer, and process validation)	A process capability ratio of 1.33 is required for entire operating range of the process parameters in injection molding of the luer lock.	The injection molding process is statistically stable, with a Cpk of 1.47 and a Ppk of 1.23 across the entire universe of possible combinations of process parameters (worst-case conditions).
Component acceptance specifications (design verification, design transfer, and process validation)	There are three relevant or critical dimensions in the luer lock that must be met at ±.001". These dimensions are length = 0.87", I.D. at 0.11", and O.D. at 0.16".	All three critical dimensions were met, including all those devices made under the worst-case conditions.
Subassembly, subsystems assembly, and testing specifications (design verification and validation, design transfer, and process validation)	Before inserting the tubing into the luer lock, the solvent will be applied, and a curing of T minutes will be allowed. The required bond strength is P pounds when stretched.	At 99% reliability and 95% confidence, a safety factor of 3 was obtained during a stress-strength test. Curing time is a design parameter that needs control in manufacturing. During the reliability test, it was realized that the connection luer lock to tubing is stronger than the elongation of the tube, which is the failure mode when stress is applied. No new hazards are added by the elongation of the tubing.
Systems assembly, final configuration, and final acceptance specifications (design verification, design transfer, and process validation)	The extra-strong IV set has the same functionality, capabilities, and dimensions that a normal IV set has. Additionally, it has a special strength and a special labeling indicating it to be extra strong.	The potential elongation of the tubing at P pounds does not jeopardize the intended use under worst-case conditions at 99% reliability/95% confidence.
Validated design with a validated manufacturing and assembly process	The DMR and the DHF are complete.	

References

Luckman, Joan, and Sorensen, Karen Creason, *Medical-Surgical Nursing*, W.B. Saunders Company, Philadelphia, 1980.

chapter five

Measuring design for Six Sigma effectiveness

Measuring the effectiveness of Six Sigma in product design and development has been one of the key topics of discussion in recent Six Sigma conferences. There is no clear consensus among Six Sigma practitioners on any one set of measures. This is because, unlike for Six Sigma-based improvement opportunities using DMAIC methodologies, measures used in the design and development efforts are not as straightforward. For example, it is easier to measure the level of improvement in cost, quality, and time using the DMAIC approach since there are established baselines. Further, measures in new product development are traditionally more focused on cycle time (time to market). Especially in the medical device business, this cycle time measure is crucial due to severe competition, changing regulations, and customer preferences. When FDA's Design Control was introduced as a requirement, most of the product development professionals feared that this metric would be negatively impacted due to increased documentation. Our observations indicate that there seems to be little to no negative impact over the past 5 years to this measure due to Design Control.

Since DFSS is seen as "another" initiative in addition to FDA's Design Control, it is possible that medical device companies may struggle on how to measure the effectiveness of DFSS implementation. Based on what we have observed, discussed, and heard there are three levels of measuring the effectiveness and efficiency of Six Sigma deployment in product design and development. They are illustrated in Figure 5.1.

These levels are based on the level of maturity of DFSS deployment within a company over time. Our opinion is that it will typically take a highly committed organization about 3 to 5 years to move from

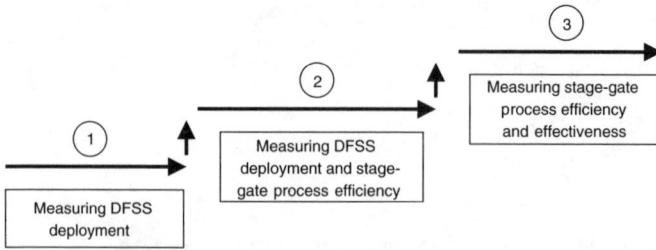

Figure 5.1 Levels of DFSS effectiveness and efficiency measurement.

Level 1 to Level 3. Let us look at each level in detail and how a medical device company that is in the early stages of DFSS deployment can reduce that timeframe.

Level 1: Measuring DFSS deployment efficiency

In the mid- to late 1990s medical device companies were in the process of implementing FDA and other regulatory bodies' Design Control requirements. At the same time, other industries were embracing Six Sigma and were able to show significant business impact. Medical device companies that were mature in their design control implementation began their DFSS journey. Since most of the industry expected their Six Sigma projects to last anywhere between six and nine months and since design control projects typically lasted longer than nine months (due to the longer product development cycle time), most of the device companies treated DFSS efforts as something outside of design control. In other words, separate projects that may or may not be related to the product development efforts were initiated and completed. Design for Six Sigma implementation was thus focused on the program efficiency and not both efficiency and effectiveness.

This level of DFSS implementation led to measures such as "number of belts trained," "number of belts certified," and "percent of total Six Sigma projects that are DFSS." While these measures certainly indicated DFSS awareness and knowledge transfer, they could not indicate the true impact of DFSS deployment. In other words, it was harder to measure business results strictly due to DFSS deployment. This approach has typically led to lack of synergy and frustration among DFSS professionals as well as product development professionals.

Table 5.1 DFSS and stage-gate process measures

	DFSS	Stage-gate
Efficiency	Number of belts trained	Time to market
	DFSS project ratio	Project budget adherence
Effectiveness	Sigma level of new products	Revenue from new products
	Percent of CTQs met	Number of customer complaints

Level 2: Measuring DFSS deployment and stage-gate process efficiency and effectiveness

In Level 2, companies measure both efficiency and effectiveness of DFSS deployment. While the efficiency approach stayed the same as Level 1, the effectiveness measure depended on the stage-gate process output for new product development. For example, when DFSS is applied to develop new products, efficiency measures such as "number of belts trained" as well as effectiveness measures such as "sigma level of design" are measured. Though this approach to measures is the first step towards DFSS integration into stage-gate process, it usually created a "two-pile" approach to measures. A medical device company might find itself inundated with measures, as shown in Table 5.1.

As a result, questions might linger about the role DFSS played in improving the effectiveness of the new product development process, especially after the implementation of FDA's Design Control guidelines.

Level 3: Measuring stage-gate process efficiency and effectiveness

In Level 3, companies measure both efficiency and effectiveness of DFSS deployment in addition to stage-gate process measures. We hope that the material presented in earlier chapters convinced the readers that all three can be rolled into one effective stage-gate process. If the medical device companies adapt our recommendation to integrate design control and DFSS into a stage-gate process for new product development, we strongly feel that there will eventually be one set of measures for the efficiency and effectiveness of the stage-gate process. This set of measures will be balanced with efficiency and effectiveness as well as leading and lagging indicators.

These should be sufficient to indicate how well DFSS deployment has enabled the new product development process. As a result, the expectations of the customers, company management, design and development professionals, and regulatory agencies to provide safe and effective medical devices on time that are also cost-effective should be realized.

chapter six

Leadership's role in ensuring success

The purpose of this chapter is to describe the role of medical device industry leaders to ensure the success of DFSS initiatives and programs in their companies. We have observed and experienced that, irrespective of the industry, most middle- and high-level executives live with misconceptions about many of the quality* and management** initiatives that have emerged in industry, with DFSS being one of them. Given this background, what can be done to ensure the success of DFSS? The strategy to ensure success starts by bringing out myths and misperceptions about DFSS and talking about reality. The DFSS leaders in the MDI must be smart, intelligent, and experienced in medical products design and must have their "feet on the ground." Credibility among those doing the design and development work and DFSS experts must be established. Proper DFSS deployment (including training, support) after these should help cure the disease of "arrogance and ignorance" as reported by the journal *Science* (Volume 269, July 1995).

The right DFSS model

Top managers in the MDI typically ask themselves if the latest management strategy or business model that they learned in their last seminar or convention could help revitalize their firms or improve their value creation. In terms of a DFSS model, the question that is typically asked is: "Should we copy the model from Motorola, or GE, or Ford Motor Company?" The most logical answer could be the GE model, since GE is the company that took Six Sigma, including DFSS,

---

* Such as TQM, TQC, TCS.
** Such as re-engineering, MBF, MBO.

to a company and management philosophy level. GE has also implemented this philosophy in its medical device group. However, like
anything else in industrial history, nothing works the same. The reality of the MDI is that just the diversity of the different medical devices
and the technologies associated with them is sufficient to require
specific design elements of an effective DFSS program. Each company
has to look at the structure of the specific segments of the industry
where it competes, decide what the underlying business strategy is
that it wants to follow (e.g., cheap commodity or premium exclusive),
and analyze what it takes to establish itself where it wants to be. From
that set of strategic requirements, the firm can then deal with its
internal parameters and constraints.

The DFSS model it follows will affect all the internal systems such
as human resources, compensation, training, management reviews,
quality, service, and so on. We are really talking about making a new
and different management program work effectively. The same
approach (as we explained in Chapter 4) that we need to establish a
relationship between requirements, parameters, and constraints for
product development is also applicable to the design and deployment
of a DFSS program. For example, there will be much more technological constraints in the development of immunoassays (e.g., assays
for hepatitis and HIV) than in the development of clinical chemistries
(e.g., assays for cholesterol and blood sugar). The DFSS leader shall
know what the state of the art really is against what is just an immature technology.

What is new about DFSS?

DFSS is aimed at designing medical devices right the first time. This
seems easy to say, but in reality it is a significant undertaking. The
many scorecards shown today may lack the accuracy and measurement resolution needed to really infer any useful conclusions. Yet
many leaders in this industry believe that implementing Six Sigma
"Motorola style" or "GE style" is an automatic infallible strategy. Let
us look back at the roadmap described in Chapters 3 and 4. Do they
really bring anything new in terms of tools to meet design and development deliverables? We do not think so. To industry scientists, engineers, and technical managers, the intentions of DFSS have always
been there. What is new, however, is the true emphasis that top
management and management consultants are giving to a
well-focused and systematic new product development approach.
The DFSS roadmap is a process that is comprised of many processes.

DFSS leaders and those responsible for its deployment should first realize that DFSS is a methodology that really enhances project management way beyond what the classical GANTT charts of Microsoft Project do. Taking this humble position is a good start for leaders in deployment, especially if the green and black belt candidates are strong and experienced engineers and scientists.

The first thing to keep in mind is that engineers and scientists are the real creators of technological breakthrough, not managers or vice presidents. When exposed to what DFSS brings to NPD, it should be very clear to the technical population that DFSS will not make them better engineers or scientists but more effective achievers in cross-functional teams. Without brilliant and dedicated scientists and engineers, any DFSS program is basically an empty shell. We advise company leaders to contemplate this reality before investing much time and resources in training skeptical belt candidates.

A benefit of adopting DFSS is the generation of systematic information in a standardized way, which simplifies answering questions like "Where are you in the project timeline?," "Why is there so much effort on one particular component?," "What are you trying to achieve by testing the software?" Also, those not directly responsible for a given specific element of design, or those who cannot understand technology design by looking at a prototype, can then read and understand the performance relationships between components and subassemblies. This systematic approach also helps in changing the attitude of lead designers or inventors who usually believe that the design is so advanced that they are the only engineer or scientist capable of understanding it. With proper leadership and support, DFSS can thus help the area of team building and knowledge sharing.

No matter what one's Six Sigma title or ranking is in the MDI, we want to emphasize that true leadership in the medical device industry starts with a reality check that MDI is not a matter of a "dog and pony show" with DFSS tools, but it is about science and engineering. The major concern should always be the real state of knowledge about the human anatomy, the healthcare procedures, the device design, its technologies, its manufacturing capabilities, its complaints, MDRs, and the threat posed to users and patients (risk management). Typical practices from commercial Six Sigma programs may not necessarily be the best setup for this industry. This is a regulated industry, where one little mistake can be fatal to either the manufacturer or the patient. To some devices, everything can be critical to quality. For example, some DFSS consultants say to design and development teams in MDI that each project should have four or five "CTQs," or

Critical-to-Quality characteristics. How can they limit what is critical? Some useful Six Sigma projects may have the purpose of improving the state of knowledge of the technology and may not "turn on" typical business leaders. DFSS projects are typically longer than classical DMAIC projects and, sometimes, everything well defined in terms of the customer will not guarantee success because of limitations to the laws of physics and other sciences. So, DFSS is not applicable to those leaders who may not be capable of thinking systematically or who may just care about results in the next six months. A DFSS program such as the one proposed in Chapters 3 and 4, with adequate balance between regulatory compliance, realistic and responsible risk management, financial benefits, and value creation, must be incorporated as part of the design and development program. Yet defining an adequate balance may also prove to be another significant undertaking, especially for new technologies.

The same way we define an organization's value based on intellectual capital rather than physical assets, management in Six Sigma companies should move to the next level; that is, to recognize that knowledge* is the fundamental raw material for modern businesses, and that there is a high value on details, real scientific principles, and strong technical expertise.

*Design, development, and scientists WIIFM***

The typical reaction from highly intelligent engineers and scientists to DFSS deployment is to see the whole program as another "flavor of the month" that will eventually disappear. These highly technical individuals also typically complain about the "state of ignorance" from their team, middle and upper management, and other cross-functional interfaces. Since DFSS and the associated systems engineering approach are precisely the best mechanisms that allow real-time understanding of the design by all those involved in an NPD project as well as all functional stakeholders, DFSS leaders must convince these highly talented individuals of these benefits, thereby alleviating their skepticism and turning them into the biggest supporters of DFSS. The following facts are aimed at answering the question, "What will it take to make DFSS deployment in MDI successful?" Fact No. 1: A big reason why Six Sigma is seated at a corporate pedestal today is because of the great publicity and the fame of the charismatic former CEO of General Electric, Jack Welch.

* Knowledge like W. Edwards Deming used to preach, "Profound Knowledge."
** What's In It For Me.

Table 6.1 2002 GE Medical Systems recalls

Product	Classification*	Reason
Angiographic x-ray systems	II	The C-arm can move at a high rate of speed, forcing the collimator into the bottom of the tabletop.
Gamma camera system (Vision T-22)	II	The collimator bolts may fail and allow the collimator to drop. Injury may result.
MR/I extremity coils	II	Extremity coil pads could catch on fire and burn patient.
Stationary x-ray system (Proteous)	II	Screws that mount the collimator interface plate may loosen and fall out, allowing the collimator to fall.
X-ray system	II	Unit may become unbalanced and tip when the C-arm assembly is extended.
Physiological ECG monitor (Solar 9500)	I	The installation of incorrect chips could result in device failure.

* FDA's designation of recall class is based on the health threat posed by the device problem. Class I indicates that a reasonable probability exists that the violative product could cause serious adverse health consequences or death. Class II is for violative products that may cause temporary or medically reversible adverse health consequences with a remote probability of serious consequences.

Source: *The Silver Sheet*, March 2003.

However, in the March 2003 issue of *The Silver Sheet*, the medical division of General Electric had six product recalls (see Table 6.1).

A close look at the reason for the recalls might make us think that these can be design flaws. Even if they are not, these recalls prove that deployment of DFSS can only minimize these occurrences but it cannot replace sound science and engineering. Many of the most significant and revolutionary medical devices invented as of today did not use any formal QFD or formalized flow-down schemes. The main source of design input came from knowing the most about the human anatomy, the medical sciences, and the drive to innovate by making something differently. The QFD and flow-down schemes were in the minds of those who came up with the ideas and the design. In order for DFSS to be successful, it must be applied in conjunction with sound engineering, science, and Design Control requirements.

Fact No. 2: The tools of Six Sigma have been around for decades. If you read a Total Quality Management (TQM) book from the late 1980s, you will see the same material that you will come across in today's Six Sigma books. As we mentioned in Chapter 1, Six Sigma comes from the top, thus ensuring that this approach is successful compared to TQM.

DFSS do's and don'ts for MDI leaders

Do's

1. Ensure that DFSS tools are positioned as enablers of the new product development process.
2. Ensure that DFSS tools complement FDA's Design Control regulations.
3. Ensure that DFSS black belts have significant and tangible involvement in the design work and that they are accountable for the success of design project(s). Make the black belts responsible for the project's outcome including regulatory requirements, schedule, COGS, and project budget.
4. Ensure that the Six Sigma organization hierarchy, with its yellow, green, and black belts, as well as master black belts and grand master black belts,* is appropriate and does not dominate the designers and developers of new products.

Don'ts

1. Don't replace quality systems regulation bureaucracy with Six Sigma bureaucracy. If you do, some of the expected results can be:
    a. R&D or NPD teams struggling to find a way to apply the Six Sigma tools. The results will be project delays and useless paperwork including scorecards with numbers that were drawn from thin air.
    b. Turf battles between quality and Six Sigma groups with the NPD teams caught in between.
    c. Don't forget DFSS when assessing potential acquisitions. Small companies and start-ups do not have the cash flows or funding to be worried about manufacturability. An approved 510(k) or PMA does not mean that the product has been developed for manufacturability. Most of these projects are Designed for Acquisition (DFA). As a result, big changes in design and "start from scratch" development

* We want to credit a medical device electrical engineer, David Yates, for this acronym.

work are possible upon technology acquisition. Those who sign the acquisition deals usually work the financials well under assumptions such as "all technical risks will be handled by our engineers."

2. Don't start a DFSS program by "baselining" NPD cycle time and budgets without considering the different complexities in each project as well as the experience and technical depth of the team members. This would be a serious case of confounding of factors and also their effects.

DFSS corollaries

1. The design and development team needs constant access to the customer. VOC should not only be assessed by marketing, but by the entire NPD team.

2. VOC, conceptual ideas, design alternatives, and all product development iterations are in a constant state of verification and validation.

3. Educate the team in the medical sciences first. Identify a minimum set of knowledge required for all team members. A project should be started with formal training in the anatomy, biochemistry, or science involved in the opportunity. A logical goal is to know as much as the doctor or healthcare specialist for whom the design is aimed.

4. The tools of Six Sigma and the entire DFSS is a more complete and robust project management program than the typical Gantt chart. However, the elements of design still require physics, chemistry, biology, and systems engineering, and those skills are needed from brilliant and energetic technical personnel.

5. Safety and efficacy are non-negotiable requirements for medical devices. Market research studies should not be wasted trying to get around these two universal requirements. Similarly, time to learn to use a device, costs, "ease of use," intuitive design, and compatibility with other equipment are "quasi" requirements, as well. Firms should not waste valuable resources during marketing research in asking these obvious questions. They should instead dig in for specific ways to satisfy the obvious requirements.

6. Ensure that some of the sessions (e.g., FMEA, QFD) to apply DFSS tools to new product design and development are facilitated by experts familiar with those DFSS tools and who have credibility within the organization.

chapter seven

Conclusion

The design and development of medical devices is an important area in the healthcare industry. Writing this book was challenging and time consuming, but it was also rewarding for us since we were able to learn and share a lot in the process. But it is our sincere hope that we were able to achieve what we set out to do. While leaders and professionals within the medical device industry ultimately make things happen, our effort in this book is to enable them with the clear understanding of how FDA's Design Control and DFSS fit together.

A review of the history of the young medical device industry would reveal the following:

1. Leadership + science & engineering → faster New Product Development (NPD)
2. Leadership + science & engineering + Design Controls → faster and better NPD

However, we sincerely believe that in order for medical device design and development to be successful (i.e., faster, better, and cheaper), the following four elements must work together at all times:

1. Leadership enhanced with a view for reality and strengthened with knowledge
2. Science and engineering
3. Design Control guidelines
4. Design for Six Sigma

If one or more areas of the above list is ignored, it may result in many unwanted consequences including but not limited to recalls, adverse events, high cost, poor market share, customer dissatisfaction, and poor employee morale.

appendix one

Glossary and acronyms

Black belt: This term is derived from the martial arts such as karate. A black belt is an expert who coaches and trains others as well as demonstrates a mastery of the art. In terms of Six Sigma, the black belts have studied and demonstrated skill in implementation of the principles, practices, and techniques of Six Sigma for maximum cost reduction or profit improvement. They are typically team leaders or internal consultants typically mastering quantitative methods such as the statistical tools and project management Six Sigma methodologies.

CAPA: FDA jargon: Corrective and Preventive Action

cGMP: Current Good Manufacturing Practices

Concept (design concept): The approximate description of the technology, scientific and engineering principles, and form of the product. Description of how the product will satisfy the (potential) customer needs and why. It is said that the NPD process can ruin a good concept, but a bad or poor concept will never be a good commercialized device.

COGS: Cost of Goods Sold

Cpk: Potential Process Capability ratio

DFx: Design for x

DFa: Design for Acquisition

DFm: Design for Manufacturing:

- Estimate the costs of manufacturing
- Reduce cost of components or raw materials
- Reduce cost of assembly or preparation
- Reduce necessary overhead
- See the effect of the above in the intended design

DMADV: Define, Measure, Analyze, Design, Verify*

DMAIC: This is a Six Sigma project methodology used when a product or its process is in existence but the output is not performing adequately (meeting the specifications). The DMAIC methodology is almost universally recognized and defined as comprising of the following five phases: Define, Measure, Analyze, Improve, and Control. In some businesses, only four phases (Measure, Analyze, Improve, and Control) are used; in this case the Define deliverables are then considered pre-work for the project or are included within the Measure phase. In other companies a second "I" has been added so it looks like DMAIIC, where the first I stands for Improve and the second I stands for Implement, while in others the first I stands for Improve and the second for Innovate. In general, the logic for the five steps is:

- Define the project goals and customer (internal and external) requirements
- Measure the process to determine current performance
- Analyze and determine the root cause(s) of the defects
- Improve the process by eliminating defect root causes
- Control future process performance

FMEA: Failure Mode and Effect Analysis is a systematic technique, using engineering or scientific knowledge, reliability, and teams (e.g., concurrent engineering) to optimize the system, process, design, product, or service. Another definition is a methodology to evaluate a system, design, process, or service for which failures (e.g., errors) can occur. Then, for each failure, the RPN (Risk Priority Number) is assigned. Action is then taken, when appropriate, aimed at minimizing the probability of failure or to minimize the effect.

GMP: Good Manufacturing Practices

GR&R: Gage Repeatability and Reproducibility

Harm: Physical injury or damage to the health of people, or damage to property or the environment (ISOAEC Guide 51:1999, definition 3.11)

Hazard: A potential source of harm (ISOAEC Guide 51:1999, definition 3.51)

---

* Verify and validate for medical devices and pharmaceuticals.

IDDOV: Similar to DMADV's roadmap: Identify and Define opportunity, Develop concepts, Optimize the design, Verify the design

Malcom Baldrige National Quality Award: This award is presented annually by the president of the United States and is designed to provide an operational definition of business excellence. The first company to win it was Motorola in 1987.

MAUDE: Manufacturer and User Facility Device Experience database, found at www.FDA.gov (FDA's data base for adverse events)

MDR: Medical Device Report

Mitigate: To cause to become less harsh or hostile (*Webster's Dictionary*)

Monte Carlo simulation: The essential characteristic of the Monte Carlo method is the use of random sampling techniques to reach a solution of the physical problem.

MSA: Measurement Systems Analysis

NPD: New Product Development

NPV: Net Present Value

PDD: Product Development Domains

Ppk: Process Performance ratio

Predicate device: According to *Webster's Dictionary*, "predicate" is to assert to be a quality, attribute, or property. In the MDI world, this term is used for premarketing submissions (e.g., 510[k]) to FDA. A 510(k) is a premarketing submission made to FDA to demonstrate that the device to be marketed is safe and effective; that is, substantially equivalent (SE) to a legally marketed device that is not subject to premarket approval (PMA). Applicants must compare their 510(k) device to one or more similar devices currently on the U.S. market and make and support their substantial equivalency claims. A legally marketed device is a device that was legally marketed prior to May 28, 1976 (preamendments device), a device that has been reclassified from Class III to Class II or I, or a device that has been found to be substantially equivalent to such a device through the 510(k) process. The legally marketed device(s) to which equivalence is drawn is known as the predicate device(s). A claim of substantial equivalence does not mean the device(s) must be identical. Substantial equivalence is established with respect to intended use, design, energy used or delivered, materials, performance, safety, effectiveness, labeling, biocompatibility, standards, and other applicable characteristics.

QC: Quality Control

QE: Quality Engineer (also known as quality systems engineer)

QFD: Quality Function Deployment

Risk: Combination of the probability of occurrence of harm and the severity of that harm (ISO/IEC Guide 51:1999, definition 3.21)

Risk analysis: The investigation of available information to identify hazards and estimate risks (ISO/IEC Guide 51:1999, definition 3.101). Risk analysis is the first step in the process of managing risk (ISO 14971). It is comprised of:

- Intended use or intended purpose identification
- Hazard identification
- Risk estimation

Risk evaluation: The second step in the process of managing risk (ISO 14971):

- Risk acceptability decisions

Risk assessment: Risk analysis + risk evaluation (ISO 14971)

Risk control: The third step in managing risk (ISO 14971). It is comprised of:

- Option analysis
- Implementation
- Mitigation
- Residual risk evaluation
- Overall risk acceptance

Risk management: Risk assessment + risk control + post-production information (ISO 14971)

Safety: Freedom from unacceptable risk of harm (ISO/IEC Guide 51:1999, definition 3.11)

SPC: Statistical Process Control

Specification: Precise, measurable detail of what the product is supposed to do. Understood in some companies as requirements or eng characteristics.

Stent: A small, lattice-shaped, metal tube that is inserted permanently into an artery. The stent helps hold open an artery so that blood can flow through it.

SWOT: Analysis of a firm's situation regarding business opportunities and challenges. The analysis includes potential internal strengths and weaknesses as well as potential external opportunities and threats.

TQC: Total Quality Control

TQM: Total Quality Management

VC: Venture Capital

VOC: Voice of the Customer

Sample Design Control and Six-Sigma plan for product and process development

DFSS deliverable in the MDI Deliverables key: I = Initiate, C = Complete, U = Update	Domain					Waterfall model and FDA guidance on Design Control
	Innovation	Customer	Functional	Design	Process	
Define the Opportunity	C					Not applicable
Identify Concepts for Further Exploration	C					Not applicable
Innovation Domain Deliverables Review and Approval	C					Not applicable
Start Customer Domain Activities		I				
Design History File (DHF)		I			C	
Project Design and Development Plan		C	U	U	U	Design and Development Plan
Marketing Strategy						
Regulatory and Clinical Strategy						
Product Development Strategy						
Process Development Strategy						
Manufacturing Strategy						
Supplier/Contract Manufacturer Management Strategy						
Safety, Human Factors, and Environmental Strategy						
Intellectual Property Assessment Strategy						
Technology Transfer Strategy						
Assess Project Risks and Develop Management Strategy		I	C	U	U	Design input
Customer/User Wants and Needs Plans for Obtaining VOC		C				
Obtain VOC: Raw and Analyzed Customer Data						
Product Description (design inputs)		C	U	U		Design input
Product Description (design outputs)		I	C	U		Design output
Customer Domain Deliverables Review and Approval		C				Design review
Start Functional Domain activities		I				
Translate the VOC into High-Level Systems Requirements		I	C	U	U	Design input
Competitive Product Benchmarking			C	U		Design input

Activity				Classification
Medical Device Functional Requirements	C	C	U	Design input
Integrated Reliability Plan	I	C	U	Design output
Preliminary Materials Selection	C	U		Design output
Sterilization Method Selection	C	C		Design output
Requirements Management Flow-Down	I	C	U	Design output
High-Level Product Risk Assessment	I	C	U	Design verification
Medical Device Report (MDR) Analysis (e.g., use the MAUDE database at www.fda.gov)		U		
Prototype or Computer Model Creation and Evaluation	C			Design output
Functional Domain Deliverables Review and Approval	C			Design review
Start Design Domain activities	I			
DFx (Safety, Human Factors, Biocompatibility, Toxicity, and Environmental) Analysis	I	C	U	Design output
Low-Level or Tier 2 Requirements	I	C	U	Design input
High-Level Product Design Selection	I	C		Design output
Component or Tier 3 Requirements		C	U	Design input
Product Design	I	C	U	Design output
Labeling and Product Insert Design	I	C	U	Design output
Packaging Design	I	C	U	Design output
Product and Packaging Design Risk Assessment Including Installation Risk (if applicable)	I	I	C	Design verification
Component and Product Parameter Optimization		I	C	Design verification
Operations and Supply Chain Strategy	I	C	U	Design output
Process Validation Strategy	I	C	U	Design transfer
Quality Strategy	I	C	U	Design output
Risk Analysis Report		I	C	Design output
Engineering Drawings		C	U	Design output
Specifications (including tolerances)		C		
Design Domain Deliverables Review and Approval	C	I	C	Design review
Start Process Domain activities	I	I		
Manufacturing Process Flow Chart	I	C	C	Design transfer

(continued)

DFSS deliverable in the MDI Deliverables key: I = Initiate, C = Complete, U = Update	Domain					Waterfall model and FDA guidance on Design Control
	Innovation	Customer	Functional	Design	Process	
Process Parameters and Operational Ranges (Characterization)				I	C	Design transfer
Component Qualification (including process capability and stability at that level)				I	C	Design verification
Subassembly/Assembly Testing Overstress/Highly Accelerated Life Testing (HALT), etc.				I	C	Design verification
System Integration (e.g., Can All the Parts Work Together?)				I	C	Design verification
Evaluate Capability of Key Processes Process FMEA Completed				I	C	Design transfer
Test Methods Validation				I	C	Design output
Quality Control and Assurance Plan					C	Design transfer
IQ/OQ Protocols and Reports				I	C	Design transfer
Software Validation Protocols and Reports				I	C	Design transfer
Design Verification Protocol and Report Including Sterility Testing				C	U	Design transfer
Clinical Release Protocol and Report for Regulatory Submission and/or Lot Release				C		
PQ Protocols and Reports Including Packaging, Sterilization, and Transportation Validation Protocol(s) and Report(s)				I	C	Design transfer

Activity	Code	Design element
Shelf Life Testing	I	Design transfer
Creation of BOM	C	Design output
Product Cost/Production Costs/ROI Calculation	C	Design output
Process/Product Documentation (Device Master Record)	C	Design output
Manufacturing Instructions, Manufacturing Traceability Sheets		
Quality Instructions		
Specifications and Drawings		
Manufacturing Scale-Up	C	Design transfer
Operator Training	C	Design transfer
Design Validation Protocol and Report	C	Design validation
Master Design Transfer Plan between Sites	C	Design transfer
Manufacturing Transfer Plan	C	Design transfer
Process Domain Deliverables Review and Approval	C	Design review
Start Post "Process Domain" activities	I	
Major Complaint Types Expected and Predicted Complaint Rates	I	Design output
Legal Clearance	I	
Release Product to Market	I	
DHF Released to Document Control	I	Design output
Lessons Learned and Project Metrics Review	I	
Project Closure		

appendix three

Further reading

Arom, Kit V., Emery, Robert W., Fonger, James D., Mack, Michael J., Robinson, M.C., Subramanian, Valavanur A., 1997, *Techniques for Minimally Invasive Direct Coronary Artery Bypass (MIDCAB) Surgery*, Philadelphia, PA: Hanley & Belfus, Inc. (Medical Publishers).

Blanchard, Susan, Bronzino, Joseph, and Enderle, John, 2000, *Introduction to Biomedical Engineering*, San Diego, CA: Academic Press.

Chowdhury, Subir, 2002, *Design for Six Sigma*, Chicago, IL: Dearborn Trade Publishing.

Creveling, C.M., Slutsky, J.L., and Antis, D., 2003, *Design for Six Sigma*, Upper Saddle River, NJ: Pearson Education, Inc. (Prentice Hall PTR).

Dhillon, B.S., 2000, *Medical Device Reliability*, Boca Raton, FL: CRC Press.

Fries, Richard C., 2001, *Handbook of Medical Device Design*, New York: Marcel Dekker.

Godfrey, A. Blanton, 2002, "Why Six Sigma?," *Quality Progress*, Milwaukee, ASQ. January 2002.

Gopalaswamy, Venky, and Justiniano, Jose M., 2003, *Practical Design Control Implementation for Medical Devices*, Boca Raton, FL: Interpharm/CRC Press.

Harry, Mikel, 2001, *Six Sigma Kowledge Design*, Palladyne Publishing.

Harry, Mikel, and Schroeder, Richard, 2000, *Six Sigma, The Breakthrough Management Strategy*, New York: Doubleday, div. of Random House, Inc.

Ireson, W.G., Coombs, Clyde F., and Moss, Richard Y., 1996, *Handbook of Reliability Engineering and Management* (2nd Edition), New York: McGraw-Hill

Juran, J.M., 1992, *Juran on Quality by Design*, New York: The Free Press (Juran Institute).

Launsby, Robert, and Huber, Charles, 2002, "Straight Talk on DFSS," *ASQ Six Sigma Forum Magazine*, Volume 1, Number 4, August.

Luckman, Joan, and Sorensen, Karen C., 1980, *Medical-Surgical Nursing*, Philadelphia, PA: W.B. Saunders Company.

Montgomery, Douglas C., 1996, *Introduction to Statistical Quality Control* (3rd Edition), New York: John Wiley & Sons.

Montgomery, Douglas C., 1997, *Design and Analysis of Experiments* (4th Edition), New York: John Wiley & Sons.

Statmatis, D.H., 1995, *Failure Mode and Effect Analysis*, Milwaukee, WI: ASQC Quality Press.

Treichler, David, Carmichael, Ronald, Kusmanoff, Antone, Lewis, John, and Berthiez, Gwendolyn, "Design for Six Sigma: 15 Lessons Learned," *Quality Progress*, January 2002.

Welch, Jack, 2001, *Jack, Straight from the Gut*, New York: Warner Business Books, Inc.

Index